The Qualified Self

The Qualified Self

Social Media and the Accounting of Everyday Life

Lee Humphreys

The MIT Press
Cambridge, Massachusetts
London, England

This book was set in Stone Serif by Westchester Publishing Services.

Library of Congress Cataloging-in-Publication Data

Names: Humphreys, Lee, author.
Title: The qualified self : social media and the accounting of everyday life / Lee Humphreys.
Description: Cambridge, MA : MIT Press, [2018] | Includes bibliographical references and index.
Identifiers: LCCN 2017038167 | ISBN 9780262037853 (hardcover : alk. paper) ISBN 978026253895-4 (paperback)
Subjects: LCSH: Information technology--Social aspects. | Social media. | Diaries--Social aspects. | Self--Social aspects. | Identity (Psychology) and mass media.
Classification: LCC HM851 .H856 2018 | DDC 302.23/1--dc23 LC record available at https://lccn.loc.gov/2017038167

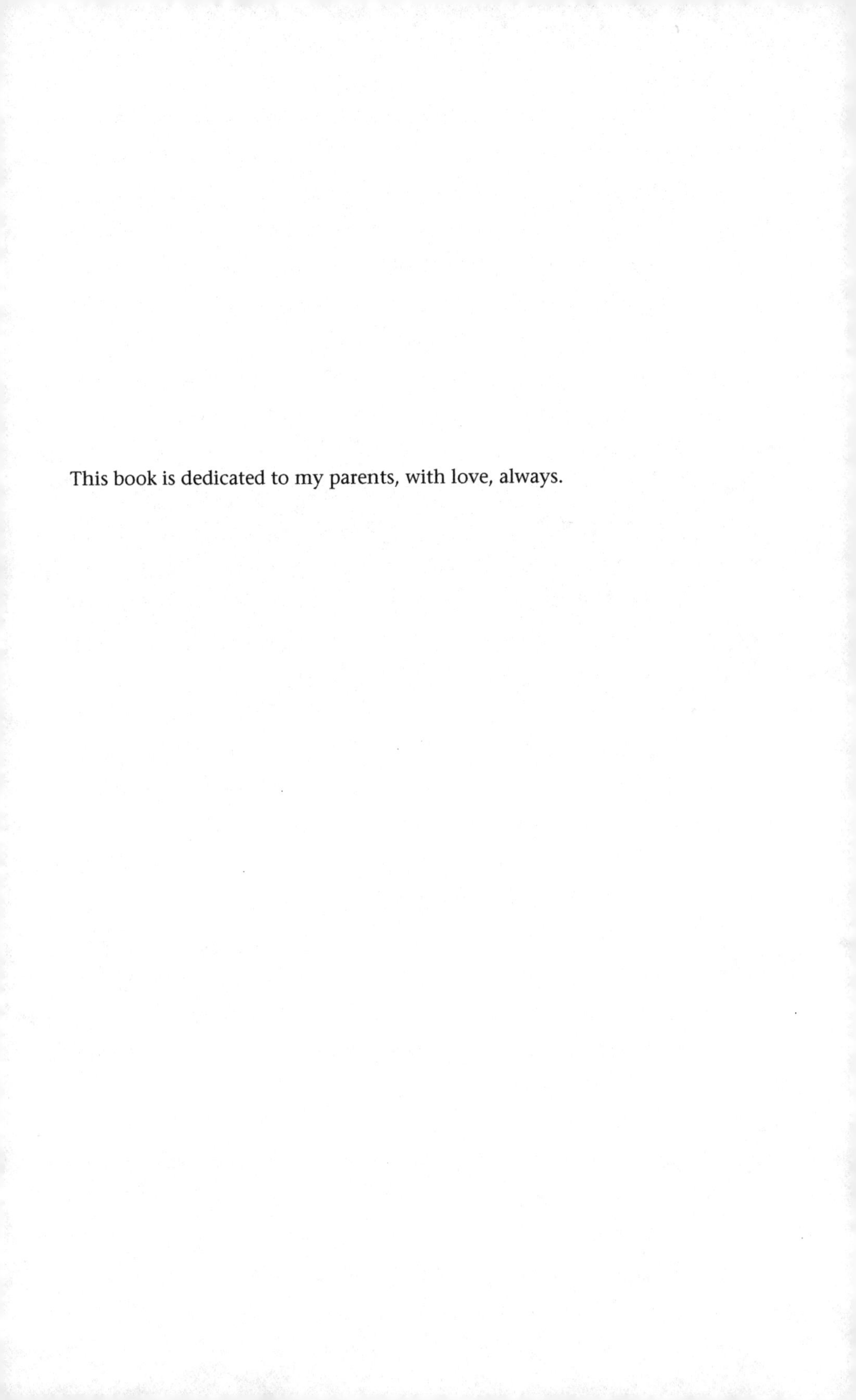

This book is dedicated to my parents, with love, always.

Contents

List of Illustrations

Preface

I've been studying mobile phone use since 2001. And there have been concerns about rude behavior and mobile phones ever since they came on the market. Early on people were concerned about the seemingly oblivious nature of people often ignoring others and speaking loudly on cell phones in public spaces due to poor reception—something they used to call cell-yell. It seemed that the less people could hear the person on the other end of their phone, the louder they spoke. "I can't hear you. Can you hear me? How about now? I think I lost you. Can you hear me now?" was a common refrain heard across the streets of America as people tried to make cell phone calls despite shoddy service. In fact, throughout much of the early 2000s, Verizon used "Can you hear me now? Good," as the tagline in their TV ads, suggesting that their cell phone coverage was much better than AT&T's. As mobile service overall improved, concerns about cell-yell have gone away, but concerns about inappropriate behavior remain.

Today, when we think about obnoxious mobile phone behavior, an image of a person taking a selfie is likely to pop up. Yes, the selfie has become the poster child for what it means to own a mobile phone. It has become a societal concern. The selfie represents our fear that mobile phones and social media to which selfies are often shared are encouraging a kind of narcissism. Indeed, there are lots of jokes made about how social media platforms are fundamentally forms of me-promotion.[1] People on Facebook say "like me," people on Twitter say "listen to me," people on LinkedIn say "hire me," people on YouTube say "watch me," people on Pinterest say "show me," people on Instagram say "look at me," people on blogger say "agree with me," and people on Tumblr say "accept me."

We are concerned that as a society we are becoming more narcissistic. And mobile and social media are contributing to this. We no longer fall in love with our reflection in the pool of water like Narcissus did. We have fallen in love with our image reflected back to us in our filtered mobile devices. Or, if we aren't concerned about our own narcissistic tendencies, then we are concerned that others are becoming more narcissistic, especially the younger Me generation[2]—"Can you remember the last time you *didn't* see a teenager taking a photo of themselves with their phone to upload to one of the myriad social media websites?"[3] Mobile and social media devices and platforms are seen as encouraging a kind of self-obsession.

But I want to suggest that mobile media are *not* the root evil contributing to a self-obsessed society. Instead, I want to convince you that mobile and social media are part of a much longer story of the ways people use media to catalog their lives and share it with others. I argue that this is a long-standing human practice and there isn't necessarily anything pathological about writing about or taking pictures of yourself and sharing them with others. In fact, it's how people have accounted for everyday life for centuries.

Where the Book Comes From

I grew up on a family farm in upstate New York. When I say family farm, I mean that it was started by my great-grandfather. His two sons took over the farm. Each of them had two sons who then took over the farm; my dad was one of the four cousins. We all lived on the same street where the farm was. I grew up in the house that my father had grown up in and where he and my mom still live today. Growing up, I was surrounded by family. My parents used to joke they never had to worry about us having a party when they were out because there was enough family around to take notice. Growing up in this environment was amazing—idyllic, really. But growing up with so much of the family around meant that what was personal or private was also collective. I knew what ailed my aunt, of course, because we would bring over dinner to help out. There was *our* family: my parents, my two brothers, and I; and then there was *the* family. What I understood to be work life blurred with home life. My dad had lunch at home every day. He didn't go into an office; he went into the fields. As kids, we would play in the fields and the barns where our dads worked. That just seemed natural to me. So when information technology tensions emerged regarding the new blurring of home and work, I found that framing troublesome. Having grown up on the farm, those divisions had never been clear. Of course, work and home lives were blurred. They always have been for family farmers. This isn't to say there weren't other divisions on the farm. On the contrary, there were very clear gender roles within the family. But my experience on the family farm never fit with notions of industrialization or the clear distinction between public and private quite in the same way as I was reading in Jürgen Habermas's or Richard Sennet's work in graduate school.[4] Part of why I embarked on this project is to acknowledge, explore, and reconcile my own personal experiences of family and work and to help me to better understand contemporary notions of public and private, collective and personal.

I am trained as a media and communication scholar. My research over the past fifteen years has primarily examined how people use mobile and social media in their everyday lives. My PhD advisor, Carolyn Marvin, helped to instill in me a historical sensibility. Therefore, I have often drawn on a variety of historical and sociological literature to better understand social processes for interaction with "new media." Fundamentally, in this book I wanted to develop a framework for thinking about how we use media, both new and old. Media accounting is my attempt to explain what people do with mobile and social media. As a historically informed scholar, I also want to suggest that media accounting is a long-standing media practice.

My new media research has been empirically grounded. I primarily do interviews and observations with mobile and social media users. I also analyze the content of messages they produce. My understanding of historical media practices primarily comes from the scholarship of historians, feminists, and literary scholars. That said, I have conducted primary research using historical texts, most often nineteenth-century North American diaries and letters, with a particular focus on the writings of women. Throughout the book, I draw on both primary and secondary sources to explicate the media accounting practices. In some cases, I have published on the data previously and note this where applicable.

Acknowledgments

This book, much like the qualified self, may be attributed to me but has really been a collective effort. I would like to thank all of the students at Cornell University who have been research assistants for me throughout the years on this project, including: Adam Agata, Adi Potashnic, Ordessia Charron, Madalyn Darnel, Betsy Distelburger, Claudia Mei, Elise Trent, Deborah Tan, Jessica Gutchess, Alexa Ortiz, Kimberly Murphy, Danielle Kellner, Lianne Bornfeld, Aaron Brody, Samantha Weisman, Yasmin Alameddine, Emma Lichtenstein, Aidan Page, Rachel Cherner, Alexa Paley, and Rebecca Wright. I had a lot of fun working with you all over the years.

I would also like to acknowledge the Department of Communication at Cornell University. All of my colleagues have been highly supportive and encouraging of this project. Special thanks to Tarleton Gillespie and Brooke Duffy, with whom I have shared various drafts of the ideas in the book. I also want to thank our New Media and Society working group, specifically Liz Newbury, Leah Scholere, Sandy Payette, Ngai Oliver Keung Chan, Caroline Jack, and Tony Liao. Thanks also go to Katherine McComas, Geri Gay, and Sahara Byrne for their wonderful conversations and encouragement.

This project also benefited greatly from my visit to Microsoft Research Cambridge in May 2013. The feedback from and conversations with Nancy Baym, Mary Gray, T. L. Taylor, and Megan Finn were very helpful at the early stages of this project.

A fair bit of work on this book happened due to the collaboration and support from Rowan Wilken. While on sabbatical in Melbourne in 2015, I benefited immensely from our flat whites and our book chats. Thanks for introducing me to the world of Australian media studies and ordinary studies. I benefited greatly from the Institute for Social Research at Swinburne University, which sponsored my stay. Thanks too to Jean Burgess, Ben Light,

Patrik Wikstrom, Tim Highfield, and the QUT Digital Media Research Center, as well as Gerard Goggin and the Department of Media at the University of Sydney, and Larissa Hjorth and Heather Horst at RMIT's Digital Ethnography Center. Further intellectual conversations and exchanges with Anthony McCosker, Julian Thomas, Jenny Kennedy, James Meese, Ramon Lobato, and Esther Milne were immensely helpful.

I have received invaluable feedback on this project from various places where I have been invited to talk. Specifically, I would like to thank Peter Decherny and the History of Material Texts Seminar series at the University of Pennsylvania and Jeremy Birnholtz and the graduate students at Northwestern University for their hosting. I also want to thank the anonymous reviewers for their helpful feedback.

Lastly, I would like to thank my family for their love and support. To my brothers, Brett and Alex, who have always been way more into books than I have. To my parents, who regularly travel down to Ithaca to see us, play games, go to the grocery store, and fold laundry. To my kids, Ruth and Charlie, who provide wonderful subjects for our media accounting. And finally, to Jeff, my partner in everything: thank you for encouraging me throughout this project. Hope you like the pictures.

1 Introduction

On January 14, 1764, Mary Vial Holyoke wrote the following in her diary: "Buried." She was writing about the death of her daughter, Polly, who had become ill only five days earlier. Her next diary entry was on January 17: "Small Pox began to spread at Boston." For forty years, Ms. Holyoke documented daily life events; typically, her entries were no longer than a line or two.[1] She recorded the births, sicknesses, and deaths of those in her family and community in Salem, Massachusetts; detailed the outbreak of small pox, snowstorms, and earthquakes; and recorded who visited her home, how food was prepared, and how much she paid for tea. She chronicled life not only for herself, but for her family and community. Her diary, like many diaries of the time, was also likely shared with friends and loved ones.

With the same brevity and abbreviation we now expect from text messages and platforms like Twitter, Elizabeth Sandwith Drinker of Philadelphia wrote in her diary on November 17, 1779:

> Stay'd at home all day—had a Beef cut up—S. Sansom spent the afternoon, S Swett, Hillory Baker Senr. & call[e]d— 60 or 70 Cabbages brought in–cloudy

Drinker wrote about the people who came to visit, about work around the house, and the weather. As a woman in the late 1700s, her job was to take care of the house, as her husband was a local merchant. She used her diary to keep track of social calls, commercial transactions, and other household activities. Her diary was a blend of her work life and social life.[2]

When people traveled, they often kept journals to document their journey and new experiences. But even the most exotic journeys involve mundane details. In the fall of 1861, twenty-year-old Ruth Bradford of Pennsylvania accompanied her father and brother on a seven-month voyage aboard a ship to China. One month into the trip, she wrote:

> Sunday, Oct. 13th I have put myself on the sick list today. The ham, eggs, and chocolate which I took for breakfast does not agree with me. Then there is a very heavy sea on, and altogether I feel a little sea sick. Think I'm done with ham and eggs forever.

Bradford, like many on social media today, documented her travel and the food she ate along the way—even what made her sick. She is not reflecting on the exotic but chronicling the mundane experiences that constitute much of travel. Indeed, travel is often only punctuated with the occasional visit to extraordinary monuments or ceremonial performances. It is through the everyday details of life, even while traveling abroad, that we see glimpses into the human spirit. Indeed, the bellyaches of life in many ways connect us to others.

What we think of today as a diary is probably a small book with a lock on it into which someone pours his or her innermost thoughts and feelings: "Dear Diary, today I fell in love with..." But this is a relatively modern notion of what a diary is. Throughout much of the eighteenth and nineteenth centuries, diaries were often written with the intention of sharing with others.[3] When young women got married and moved away from their parents, they would send their diaries back home as a way of maintaining connections with family. It was not uncommon for travelers to send their diaries home so their families could read them and know what they were up to while away. When distant family would come to visit, it was common to have them read the diary or read through it together as a means of catching up. Historically, diaries, particularly those of women, chronicled everyday life activities and events of the household and community—who was born, who got sick, who died, who got married, who visited, what was planted, what was made, and so on. Women played an important role in the social chronicling of family and community events. Sometimes these events were tragic, sometimes they were mundane, but all were recorded in these diaries and, most importantly, they were shared with others.

What's New Is Old

Diaries like those of Holyoke, Drinker, and Bradshaw provide a lens that helps us historically situate contemporary social media. Today, people use social media to document and share their lives. People share on Facebook when they are moving or have had a baby. They tweet about what they just saw on TV or what they had for breakfast. They post photos on Instagram of beautiful sunsets on their way home from work or their dog looking adorable. They post from or check in at baseball games or concerts. These are all ways of using social media to document what people are doing and sharing this with others. These are things people used to do (and still do) with diaries, photo albums, and scrapbooks. People are using new media in similar ways to how people used old media.

So often novel media are compared with their immediate predecessors. In the case of social media, new platforms like YouTube and Facebook were contrasted with either Web 1.0 or television. Much of what was new about social media was the ability to let ordinary citizens have a platform from which to speak to the wider world or the ability to share content among peer networks. But diaries like Drinker's reveal how what seems new may actually be old. User-generated content only seems novel when contrasted with mid- to late twentieth-century understandings of media as broadcast mass media. By comparing and contrasting media across longer historical periods, we can begin to understand some of the critical tensions surrounding social media differently. This book develops a framework for understanding the ways that people use media, broadly defined, to chronicle their lives and share it with others.[4]

Old and New Affordances

One way that we compare the differences between old and new media is through their technological affordances. The concept of affordances helps us to see how the characteristics of media technologies can invite us to use media in particular ways.[5] Understanding the materiality of diaries has informed my understanding of technological affordances more broadly. We can see the shrinking of a diary page as a technology that invites us to write in particular ways. Of course, we could devote a whole journal to the chronicling of one day, but we don't. We take the smaller pages as an invitation to write less. Similarly, when dates were printed on journal pages, we took it as an opportunity to write about what happened on those days, even though we did not necessarily have to. When they started printing horizontal lines on the pages, we took it as an opportunity to write between those lines, even though there was nothing to stop us from writing perpendicularly to the lines. We take these characteristics of the technology as invitations to use the media in certain ways and always have.

Diaries from previous centuries have ranged in size and shape, but most are relatively small—and portable. In the mid-nineteenth century, advances in the production of paper facilitated the production of smaller diaries. Historian Molly McCarthy describes these "pocket diaries" as about two inches by four inches in size and, as the name suggests, were intended to be tucked into pockets or waistbands.[6] The size of the journal afforded mobility such that people could carry it with them and jot things down in it throughout the day rather than waiting until the end of the day. It became a "real-time" form of chronicling. The size of the pocket diaries also constrained the length of an entry to a sentence or two. Other nineteenth-century diaries were larger, but even these were designed so that there were only a couple of lines per date on which to write.[7] There would typically be three dates printed on each page of the diary. These diaries were also little books of mostly lined paper that could be easily opened and shared with others.

The material characteristics of diaries (that is, small, mobile, space-constrained) are not all that different from the character limit on Twitter. The 140-character limit (now increased to 280) was originally put in place to ensure that Twitter was cross-platform; it allowed users to send and receive tweets via the web or SMS, which limited the length of a text message to 160 characters before breaking it into two (or more) messages. But beyond this technological reason, early adopters seemed to enjoy the character limit. Some of the earliest adopters of Twitter were bloggers, who likely saw the limit as a welcomed constraint, much like diarists of the mid-nineteenth century. As McCarthy writes,

> The space afforded by the pocket diary may have been limited, but it saved journalists with only minutes to spare from having to write long entries. And diarists appeared thankful for both the opportunity pocket diaries offered as well as the limitations they imposed.[8]

Technological constraints can be welcomed because they delimit what we might otherwise feel obliged to do. It's not always about what we can do with a technology; sometimes its value is in what we *cannot* do with it.

For many of us in the developed world, we first came online through a computer that was plugged into the wall. Now, however, we typically access the Internet via mobile devices. For many people in the world today, the first time they access the Internet will be on a mobile phone.[9] Thus our experience of the Internet and social media is increasingly mobile, just like diaries.

Ordinary Culture

My intention with this book is to highlight one of the ordinary ways that people have used and continue to use media in their everyday lives. But the ordinary can be deceiving. Often, we miss it altogether. We don't pay attention to it. It goes overlooked in the shadow of the moment. I believe this has happened to a degree with social media research. Much of the new media literature published in the past ten to fifteen years examines its extraordinariness, explores how networked media are different and new and how they are changing social interactions,[10] and identifies structures of publicness.[11] The role of social media in revolutions, political campaigns, and responses to natural disasters has become a prominent lens through which we study and explore social media.[12] Even the role of social media around events like Eurovision or the Oscars has been an area of research.[13]

That being said, these events do not happen every day. So what happens in between awards ceremonies and natural disasters? What do people do with mobile and social media when they wake up in the morning, when they wait for the train, when they're bored at work? I argue that we need to understand the everyday routines and practices around social media for two main reasons. First, if people are going to use these technologies for major events, they need to be familiar and active on these technologies before and after such events. To study the discussion of Eurovision or the Arab Spring on Twitter means to study the phenomenon itself. But to study the discussions of the events, people had to know how to use the platform prior to the event. Part of how we know that there was a smallpox epidemic in Boston in the late eighteenth century is because people wrote about it in their diaries. These are the same diaries in which they wrote about planting corn and who visited. It was because the everyday was being recorded regularly that the eventful moments could be captured in these diaries as well. If we are to study Arab Spring, for example, through Twitter, then mobile phone adoption and use—the mode through which most people access Twitter[14]—had to be in place before the elections and demonstrations. If people are to capture police brutality on their camera phones, like they did in the case of Eric Garner, who died of a heart attack while New York City police were trying to arrest him in 2014, then people need to have them in their pockets to begin with. So, these events raise the question of what happened before: What were people doing with their smartphones in their pockets? Why were people writing down daily events and news in their diaries? What is so important about mobile media that people keep them so near to their bodies and use them every day? What makes media become so ordinary that they can help us capture the extraordinary when it happens in front of us?

The second reason we need to understand the everyday aspects of social media is because the ordinary can represent broader social values and systems that shape the human condition. In this case, it is essential to understand what I mean by ordinary. Ben Highmore writes:

> Ordinariness is a process (like habit) where things (practices, feelings, conditions, and so on) pass from unusual to usual, from irregular to regular, and can move the other way (what was an ordinary part of my life is no more). There is always the "being ordinary" but there is also the "becoming ordinary."[15]

Much of my research has focused on media "becoming ordinary"—that is, the domestication and emergence of social norms surrounding mobile and social media use.[16] Carolyn Marvin argues that it is in the early stages of technology development and adoption that the tacit understandings and assumptions about media are made explicit.[17] The questions "What is this new media?" and "What is it doing to us?" are both actively discussed by people face to face as well in through other forms of media, such as online discussion forums, newspaper articles, and various how-to guides. My research over the last fifteen years has involved talking to people about their mobile and social media use. It is much easier to talk to people about their social media use compared with something like washing machine use because their social media use is still new; therefore, they are still working things out. A common question I ask is, "How would you describe *X* to someone who has never used it before?" That question works because my participants can imagine that someone might actually not be familiar with a particular platform. In fact, they might have had to describe the mobile app that I'm studying to someone they know. The question becomes much harder when you imagine that everyone knows what *X* is. It is much trickier to describe a TV because one imagines that most people have some experience with it. TVs have already become ordinary for many of us today, but mobile and social media are still very much in the process of becoming ordinary. We can still imagine those who might not use them.

Ben Highmore argues that the ordinary is also very *connective*. The ordinary unites us in many ways. Often, we might not recognize the ordinary because we are all doing it. We all eat breakfast (or should) every day. We all get sick. Of course, not all of us actually do these things. Not everyone can afford to have breakfast every day or has access to food. But when we do acknowledge and document our ordinary life and share it with others, we can be brought closer together. Knowing someone's ordinary routines can be a sign of intimacy. Connective rhythms of the quotidian, shared expectations, or understandings of daily routines may be tacit and normative, but they reflect a togetherness of ordinary culture.

The ordinary, however, is also highly *contextual*. What is extraordinary to one person may be ordinary to another. For some, it may have felt extraordinary to use a social media platform for the first time. In our study of tweets, we found many people tweeting things like, "hello twitterverse, I'm here now," or "just joined, what now?" When someone is new to a social media platform, they might not know where the buttons are or where to read or how to write in a manner that's typical for that platform. They do not know the norms, that is, the prescriptive collective ways of using a social media platform. It is often extraordinary to them.

The contextuality of ordinariness with regard to social media is not solely based on the amount of experience with a platform. What is ordinary about social media use among some social circles may not transfer to other circles. What is ordinary social media use for a teen might not be for an adult.[18] What one person's Twitter or Facebook feed looks like may be very different from someone else's; some call this a filter bubble.[19] Our understanding of what is ordinary on a particular platform is shaped by what people see on that platform, which is particular to that individual. The collective and contextual nature of everyday media use is important to explore and understand because it is in those nuances that we experience each other's humanity and we recognize ourselves as both unique to and part of a social collective.

My intention with this book is to identify the ordinary within new media practices by comparing them with historical media practices. Although it might seem new and extraordinary that people are tweeting what they had for breakfast, when put into context with historical diary practices, it reveals the ordinariness of the act itself. When we look across media over time, we see patterns of how people are incorporating media into their everyday lives. By focusing on what people *do* with media rather than on the technology itself, we can see similarities otherwise obscured by the newness of the platform.

Although this book is fundamentally about media practices, I do assume that the materiality and affordances of the platform matter. Being highly influenced by science and technology studies (STS), I also approach media in the book as fundamentally "media technologies."[20] To bring an STS framework to media means to understand how technologies are part of and embedded within sociotechnical systems. In particular, I draw on the social construction of technology framework to focus on the importance of *users* in shaping how we come to understand what a certain technology does, a move common within STS.[21]

Before moving too far forward, it seems helpful to define some of the terms I use throughout the book. For example, I use "media" both broadly and inclusively. I do not mean merely electronic or digital media. Media has been defined as all the channels and means through which people share information that is not face to face[22] or the tools people use to communicate with others about a shared reality.[23] For the purposes of the book, I draw on both definitions and suggest that media are those tools and channels that connect people across time and space and allow for the sharing of meaning. I use the term "meaning" rather than information or reality because I want to specifically highlight the identity expression and community building that occur through media.

It might seem curious to suggest that diaries, scrapbooks, and photo albums should be thought of as media. Although diaries have long been used as primary sources in such disciplines as literary studies and history, seldom have communication and media studies scholars found diaries to be within their purview. Historical communication scholars have focused on broadcast and electronic media such as radio,[24] television,[25] and the telegraph.[26] That said, I am not the first to argue that scrapbooks, portraiture, and snapshot photography fall under the purview of media studies, nor am I the first to suggest these are historical predecessors to contemporary social media.[27] But diaries have seldom been considered by communication scholars as communication or media. Instead, they have been relegated to methodology for studying media use.[28] For example, diaries become a way to track the TV shows that an audience member watches. However, diaries are not just a genre of personal writing but historically have been tools for collective and shared meaning making, the social nature of which is fundamental to understanding diaries as media.

Contemporary networked and social media have extended our definitions of media such that we can see beyond institutionalized broadcast media like newspapers, magazines, radio, film, and television. Media today include interpersonal forms of mediated communication,[29] and with this broader definition of media we can apply media frameworks to nonelectronic forms of mediated communication including diaries, scrapbooks, and photo albums.

Media Accounting

Throughout this book, I develop a theory of media accounting. *Media accounting* can be described as the media practices that allow us to document our lives and the world around us, which can then be presented back to ourselves or others. I draw on Nick Couldry's concept of "media practice" to describe the activities, uses, structures, and conceptualizations of and surrounding media.[30] A practice-oriented approach to media allows us to see similarities despite the differences in platforms or technologies. Identifying the key practices of media accounting focuses our attention on what people do with the media rather than focusing just on the media technologies themselves. This is not to say that the technologies themselves are unimportant, but, drawing on the social construction of technology, I ask, "What are the needs and understanding that people bring to the technology that shape its usage?"

Media accounting involves the creation, circulation, and consumption of *media traces*. A trace is the mark or vestige remaining and indicating the former presence, existence, or action of something. Therefore, media traces are vestiges or marks that indicate our presence, existence, or action through media, that is, those tools and channels that connect people across time and space and allow for the sharing of meaning. At first, media traces might seem like digital footprints, defined as the record of online activities that people may or may not be aware of creating as they use the Internet.[31] But they are quite distinct: whereas media traces are constructed by and visible to the individuals who create them, digital footprints include both purposeful postings online as well as IP addresses, clickstream, and authorship data—data about people and their behaviors that many users are not aware exist.[32] Moreover, media traces are not necessarily digital. Media traces are the texts, videos, and images created by people in their course of documenting their lives, what they do, where they go, who they are—and sharing this with others.

Media traces are essential to what it means to *document* through media accounting. Lisa Gitelman argues that to document means to know through showing.[33] Therefore, media accounting is more likely to occur through certain media than others. Diaries, journals, scrapbooks, photo albums, videos, and social media posts all privilege the showable trace, whereas a landline phone or CB radio do not create showable records of the content exchanged to be revisited or shared at a later time. Media traces are central to the practice of media accounting. Three aspects of media accounting help us to understand the practices of media accounting: an account, accounting, and accountability.

An Account

The term *account* suggests that a collection of media traces created through social media is tied to identity. An account is associated with an offline identity; think of a bank account or a store credit card. It often is individual, but it may be collective. If we think about bank accounts, individuals can have an account but they can also share a joint account with partners or family members. Organizations or groups can also have accounts. But all accounts are linked to some kind of identity.

Most social media platforms require users to create an account, typically requiring a username, password, and email address at minimum.[34] This is part of the process of tying an account to an identity. Sometimes fictional social media accounts are set up,[35] and sometimes there are accounts for bots on social media platforms.[36] But the accounts are created by someone or someones and therefore are still tied to some kind of identity.

Historically, the connection of identity to media accounts has varied in formality. Often, we think of diaries, scrapbooks, and photo albums as belonging to a person. Indeed, the front pages of a journal frequently include an empty line on which the diarist is invited to write his or her name, along with the date if it isn't already printed. Prompts such as "This diary belongs to ______" or "If found, please return to ______" are also formal ways to tie an identity to media accounts. Other times, diarists would just write their names in the front pages.

Beyond labeling the diary with one's name or the social media profile with one's username, the content of media accounts is closely associated with the identity of the user as well. Who is described, mentioned, revealed, and photographed is another important means of linking media traces with identities. Although an account may be tied to *an* identity, the content of media accounts is seldom confined to singular identities. Authors frequently include themselves within the content of their media traces, but they also include others. Various kinds of social relations—including friends, kin, enemies, frenemies, colleagues, followers, lovers, and love interests—make their way into the content of our media accounting. Their identities as well as our own are intertwined in our media traces.

All media accounting is tied to the identities of both the creators as well as the subjects, which may or may not be one and the same. Media accounting can be done not only collectively and collaboratively but also on behalf of others. For example, archivists of various social groups or organizations create media accounts on behalf of the group. Similarly, blogs can be associated with a single author or with multiple collaborators. It was common in the mid-nineteenth century for women to keep an account for the household—who visited, how much she paid for flour, who died.[37] Starting in the late nineteenth century, mothers kept baby books for their children, documenting social and developmental milestones.[38] In the late twentieth century, it was often women who played the role of family historian, documenting and creating traces of their ancestral past.[39]

The term *account* also suggests a kind of subjectivity. For someone to give their account means they give their perspective on an event. It does not mean to encapsulate the entirety of the event but merely their version of it. Such subjectivity therefore conveys a partiality or incompleteness of one's account. This is true for media accounting practices as well. All media accounting practices are subjective and incomplete, though they may be vast. We can never document all of lived experience. Although we can describe parts of it, photograph it, and even record it on video, these media traces are always already incomplete.

Like other forms of identity work, media accounting can represent strategic presentations of self.[40] The subjectivity of media accounting means that one's version or take on something is both situated and performative. We know only what we have experienced, but we are also aware of our experiences as part of our various identity performances. As such, our media accounting captures both our versions of the world as well as our aspirations for it and for ourselves. The strategic nature of our media accounting is also related to the evaluative nature of accounting.

Economic sociologist David Stark ties the term *account* to organizational criteria for the evaluation of worth. He writes, "We keep accounts and we give accounts, and, most importantly, we can be called to account for our actions. It is always within accounts that we 'size up the situation.'"[41] In particular, Stark describes the contemporary mechanism of organizational performance evaluation as horizontal accountability rather than just hierarchical evaluation. People feel accountable not just to their bosses but also to their colleagues. Employees' sense of worth within the organization is based on the evaluation of those horizontal to them, not just above them. Outside of an organizational context, media accounting can thus be thought of as an everyday way that people can evaluate the worth of actions, behaviors, events, and individuals.

The concept of an account also conveys the point that media accounting involves the collection of media traces. As a bank account is made up a series of documented financial transactions between parties, media accounting involves not just singular media traces but a collection or aggregation of traces indicating our presence, existence, and action. Sometimes media traces are referred to by their mode, such as text, image or video; sometimes they are referred to in relation to their platform, such as tweet, blog post, diary entry, photograph, Instagram, updates, check-ins, or vlog post. All are traces indicating our presence, existence, and action, and all involve connecting people across time and space to share meaning.

Accounting

Media accounting is also fundamentally about the practices of documenting, chronicling, and cataloging. Accounting is the action or process of reckoning. It can involve counting and enumerating and be aggregative and transactive. Accounting provides evidence and explanation. Media accounting, therefore, is the process of reckoning or providing evidence for and explanation of our presence, existence, and actions through media.

Accounting is colloquially associated with financial information or economic entities, but we have long used media to reckon other kinds of information in our lives. In the United States, we have used accounting to keep track of household information since the eighteenth century. We have used travel journals to keep track of how far we've traveled and where we've been. But there are others forms of information that we have reckoned with media accounting. We use our media traces to see the path of where we've been and how we got to where we are. From baby books to photo albums, we document the changes in our lives in real time, or at least relatively defined as the "near past." When we look back on these books, we see trends and changes that we may not be able to see in our lived experiences. Accounting through media is fundamentally about chronicling various aspects of our lives so that we can remember, relive, recount, reconcile, and reckon at future points.

Enumerating activities, events, and experiences can be part of what we do when we engage in media accounting. Enumeration is easy to see in the contemporary social media environment, where our profiles include the number of friends or followers we have or how many tweets we have posted. But we have long understood ourselves by various numbers: how many yarns of wool we spun, how many bales of hay we loaded, the age at which we took our first step. Accounting is a way of seeing patterns and gaining insights that might not otherwise be identified in our lived experience. By documenting activities or events through media, we can aggregate information over a season, a year, a lifetime—information that can provide explanation of our lives and our livelihoods.

Of course, media accounting is not only about enumeration. The meaning making that comes from the aggregation of information over time is indeed far more important to the accounting process than any single metricized trace. When we look back at ourselves in photos, we might see our hair differently than we did at the time the picture was taken. We see our families differently as they grow and change. We see ourselves in our familial traces. Media traces allow us to gain distance and reflection on experiences and behaviors, but the *collection* of media traces allows us to gain additional insight that an individual trace may not convey. The whole is greater than the sum of its parts. A photo album, a diary, a Facebook timeline, or a Twitter stream conveys far more information about people's existence, presence, and actions than their singular media traces do. As such, media accounting tells a story, conveys information, and reveals explanations. Media accounting is an important way through which we come to understand processes and changes, both about ourselves and about others.

But these aggregated media traces also help others understand *us*. Various social media and telecommunications companies and governments, as well as their partners, can analyze, study, and examine our traces in order to better serve, protect, and market to us. Alison Hearn reminds us that our social media traces make up the reputation economy, which drives online business.[42] Diaries, scrapbooks, and photo albums can be passed down within a family, are collected by history buffs, or become part of archives that combine media accountings. These traces are evidence of behavior, relationships, and affinities that can be read by others. The aggregation of such traces can reveal trends across historical periods or segments of the population.

Accountability

Media accounting also suggests the importance of accountability. Accountability is fundamentally about one's liability to account for and answer for one's conduct. There are three primary ways to understand accountability within media accounting. First, we are accountable to others for the traces we create. Second, we are accountable for the traces created about us by others. Third, we are accountable to others for the traces they create about themselves.

To say that we are accountable for the media traces we create is related to the evidentiary nature of media accounting. If I write a diary entry or a tweet that says someone died, I am accountable for the veracity of that information. Its existence is evidence that the death occurred—not fact, just evidence. Because I wrote it, I am accountable for it. If I write that I killed someone, then I am accountable for both the fact that I wrote it and the act itself. The distinction of accountability for both the content of the trace as well as the creation of the trace shapes the practice of media accounting.

Indeed, the accountability of media accounting is often persuasive. The earliest of diary keeping was for religious purposes, where diary writing was seen as a way to encourage pious behaviors.[43] During the mid- to late eighteenth century, diary writing was often seen as a way to reflect on and shape our behaviors. Historian Jane Hunter writes:

> Parents and authorities promoted diary-writing among girls as an effort to contain selfishness and encourage conformity to social expectations. Like the Catholic confessionals described by Foucault, diary-writing was an internalized discipline of the self.[44]

We become accountable not just to others for the media traces we create, but to ourselves as well. Thus, media accounting can also be a means of shaping, influencing, and persuading future behaviors, actions, and thoughts.

Media accounting is presentist not only in its recording or chronicling but also in its reading. Past traces are experienced in the present and can dramatically influence our present understanding of self and others. As such, for better or worse people are responsible for their creation of media traces and traces about them sometimes long after the trace was created.

We are also accountable for our traces over time—traces we created in our youth or at a different time in our lives. For example, students are constantly reminded to delete Facebook photos of themselves holding red Solo cups before applying for college or going on the job market. While our beliefs or actions depicted in these traces may have been normative at the time, situations change, and interpretations of these can traces change as well.

Because media accounting is tied to an identity, it is also tied to roles and responsibilities. There are normative expectations that go along with various social roles such as being a good wife, mother, daughter, professor, friend, citizen, etc. In our media accounting, we are accountable to the same social roles. Therefore, we may strategically choose to create traces that reinforce or reflect these social roles. As a mother, I feel pressure to chronicle the lives of my children, to take photos of them on the first day of school or with their birthday cakes. The only thing that's changed is whether I post it on Facebook or put it in an album. But the middle-class norm to document the event or milestone has existed for a long time.[45]

We are also accountable for the media accounting involving us. Because media accounting is social in nature, we are often figured into the traces of others. Sometimes media traces are explicitly crafted about us *by* others; for example, baby books. Other times we figure more tangentially into the traces of others. When we visit someone or go to a concert or party with a friend who then posts a photo or writes or tweets about it, they are writing about themselves but it involves us. Sometimes traces are collectively made, such as school class photos, which are the traces of the school, the teachers, and each of the students. If a teenager posts that she went to a bar with a friend who is underage, the underage friend must defend the trace as much as the original poster. Even though we may not be the one creating the trace, we are accountable for our presence in the traces of others.

Last, we are accountable for hearing, watching, and reading the media traces of others. Sometimes media accounting is a highly dialogic practice. We write with an audience in mind or with an expected response. When I wrote on my blog that I went to a special restaurant, I'm assuming that my mom was reading and would find it interesting because she loves food and me. She knows I kept this blog about our sabbatical experience in part to keep her updated on things. As such, she has a responsibility to read it. I think she genuinely wanted to read it, but there were expectations that she would read it as well.

When someone on Facebook posts good news, such as getting married, or has a birthday, we are accountable for receiving that information—not just by reading it but by bearing witness, that is, by somehow communicating that we are receiving that information in the ritual of receipt. Sometimes this is in the form of a response or a question—"Way to go! Where was that photo taken?" Sometimes, we subtly incorporate the information into future conversations: "You've been there, right?" Responses can be as minimal as clicking a Like button. Sometimes the ritual of receipt is a very physical act of receiving media accounting, such as sitting down with someone at the table to look through a scrapbook with them. Parents in the nineteenth century would read their children's diaries aloud at the end of the day.[46] It was very common throughout the mid- and late twentieth century to sit through a friend or family member's projected slide show of trips or holidays.[47] Throughout the 1970s and 1980s, slide projectors were a popular means of sharing media traces. I still remember sitting through my grandfather's slides of his trip to China—lots of buildings and hills, few people. It lasted an hour—a long presentation, especially for a seven-year-old like myself. But there was a social obligation to watch the slideshow as my grandfather narrated. There was a social obligation to bear witness to the media traces he created of his journey and experiences. We had an accountability to do so. The rituals of receipt may vary, but there is accountability to receive the media accounting of others. These rituals of media accounting reinforce our social bonds.

Our accountability for the media traces we consume has been tested when people have posted suicidal messages on social media. How accountable are we for the media accounting of others? In short, it varies depending on our relationship with the poster, the media, and the context. But the fact that various actors are accountable plays an important role in the value of media accounting and what makes it a powerful communicative act.

What Media Accounting Is Not

I do not want to suggest that all mobile and social media are reflected in media accounting. Indeed, there are a plethora of contexts and uses of mobile and social media and therefore no one theory can account for the complexity of practices that surround such media. Media accounting is not about the creative vernacular, to use Jean Burgess's term, prevalent on mobile and social media.[48] It is not about the wonderfully creative videos that artists, media producers, and aspiring influencers post.[49] Media accounting is not about political organizing or affective publics,[50] even though these also describe mobile and social media. Media accounting is not about how companies and corporations use social media to advertise their products and services or build their brands,[51] though advertising has been integral to and has supported media accounting practices since the early twentieth century.

Instead, this book explores a discrete set of practices that various forms of media—digital and analogue, textual and visual, social and seemingly antisocial—have been used for. Just as there are many uses of social media, so too are there many uses of diaries and photo albums and scrapbooks. Looking across media to see patterns of communicative practice is not meant to belittle the significant differences in media platforms as distinct sociotechnical systems situated within distinct historical periods. Rather, it is to highlight long-standing human needs to use media to record and share our versions of the world as a means of making sense of the world and our place in it.

The Qualified Self

I want to suggest that the sense of self that emerges from media accounting can be understood as the qualified self. As we create media traces of ourselves in writing, images, audio, and video, we create representations of ourselves to be consumed. These media can be read back to ourselves or others. It is through the consumption of these media traces that we come to understand ourselves and others, sometimes in new ways. We can hear and see things in media traces about ourselves and others that we might not have noticed in real time. When we look at photos of ourselves or read what we have previously written about ourselves, we can engage, relive, and scrutinize ourselves from perspectives different from our lived experiences. The qualified self is the understanding of ourselves that emerges from creating and reengaging with media traces. The qualified self can be broken down into three aspects: quality, qualify, and qualification.

Quality

Media accounting conveys one's character, disposition or nature, that is, one's qualities. When we engage in media accounting we depict qualities about ourselves and others. The choice of what we decide to create a trace about conveys our attributes to others. For example, I might post about something I read in the *New York Times* but I might not post something I read or (more accurately) looked at in *People* magazine. I read them both, but only chose to create a trace about one. Qualities can be implicitly or explicitly communicated as we create media traces about ourselves and others. Qualities can be given or given off.[52] Character or disposition can be strategically created in our traces or naively revealed in our media accounting. Regardless of intention or explicitness, a qualified self is defined by the qualities and attributes of people and their experiences as evidenced by their media accounting.

It is important to recognize that quality often conveys something desirable. A quality is often something valued, a virtue. The qualified self, therefore, often focuses on people's qualities. The qualified self conveys strengths and virtues. Some might call this a positivity bias, that is, a tendency of humans to communicate more positively. A positivity bias has been shown to exist not just on social media,[53] but in language more broadly. Peter Dodds and colleagues found that across languages and cultures positive words are "more prevalent, more meaningful, more diversely used, and more readily learned."[54] The qualified self similarly belies a positivity bias with regard to the qualities that define the qualified self in media accounting.

Qualify

The qualified self can also be understood as qualifying. To qualify can mean to describe or to designate in a particular way. The qualified self is therefore a described self, a characterized self. The aggregation of descriptions, of our media traces, and of the media traces of others that feature us, convey a particular version of who we are, a qualified version. However, to qualify also means to modify or moderate. In this way, the qualified self is a modified version of the self. We cannot possibly create traces of everything in our lives. The qualified self is a modified and a selective version of the self represented in media traces to be interpreted by ourselves and others.

The qualified self is a described self that must be interpreted, not just analyzed. Harry Wolcott distinguishes analysis from interpretation through the results of the process.[55] Analysis results in patterns, but interpretation results in meaning. The qualified self privileges interpretation and reckoning over analysis. We find meaning in the juxtaposition of photos of the first and last day of school because they are evidence not only of our children's physical changes over the school year but also of the changing emotions elicited, from anticipation to accomplishment. Our media traces of who we are and what we do circulate within modes of interpretation and meaning making.

Qualification

The qualified self is also made up of our qualifications. Media traces are evidence of who we are and what we've done and therefore communicate our accomplishments. The qualified self is an aggregation not just of behaviors and experiences but of achievements and milestones. The qualified self conveys our qualifications for our various social roles. When potential employers and romantic interests search for us, our media traces help them understand who we are and what we can do. Media traces of what we have done can become qualifications for future experiences as media accounting circulates.

The Self

Notions of *self* are essential to the qualified self. On a theoretical level, the term "self" implies both a subject and an object. A person engages in media accounting as a subject—that is, as a creator of traces—but also experiences oneself as an object through media accounting by seeing oneself in the traces created by oneself or others. The qualified self is situated within sociological understandings of selfhood which position the self as essentially social. Sociologist George Herbert Mead writes:

The self is something which has a development; it is not initially there, at birth, but arises in the process of social experience and activity, that is, develops in the given individual as a result of his relations to that process as a whole and to other individuals within that process.[56]

For Mead, communication is the process by which individuals may become objects to themselves. Communication is an essential experience to the development of the self. Media accounting is therefore part of this social development of the qualified self. It is in and through media accounting that the qualified self develops both as a subject and an object. Moreover, the qualified self, to quote Mead again, "arises in social experience."[57] This means the qualified self is a social self and media accounting is fundamentally a social experience.

The qualified self is also based on feminist understandings of the self. Throughout the late eighteenth century into the nineteenth century, women in New England did not merely write about themselves.[58] Their familial and social relations figured prominently into their identity, their selfhood, and their media traces. Peter Heehs's extensive history of the self is fundamentally the history of a masculine self.[59] Famous men wrote about themselves with a great sense of history, but the women in diaries and scrapbooks more often created traces not just about themselves but about others as well. The qualified self draws on feminist scholarship, which draws our attention to the everyday wives and daughters who track themselves, their families, and their communities in their pages and posts.[60] The qualified self is not necessarily a feminized form, but takes on a feminist logic that defines the self in relation to others.

The Qualified and Quantified Self

The term "qualified self" evokes the quantified self movement. This is a movement that has gained great traction within the last twenty years with the rise of networked and digital technologies. In her book, *The Quantified Self,* Deborah Lupton writes:

While the quantified self overtly refers to using numbers as means of monitoring and measuring elements of everyday life and embodiment, it can be interpreted more broadly as an ethos and apparatus of practices that has gathered momentum in this era of mobile and wearable digital devices and of increasing sensor-saturated physical environments.[61]

Lupton argues that the quantified self is broader than just the enumeration of behaviors. The quantified self is part of a lay movement to generate knowledge about bodies within a Foucauldian framework of self-knowledge and discipline. As a scholar of the sociology of health, Lupton frames the quantified self as a long-standing practice of self-tracking, which digital culture has made increasingly prominent as well as contested with regard to privacy, embodiment, surveillance, and knowledge production.

The term "quantified self," however, is contested. Gina Neff and Dawn Nafus argue that the Quantified Self is a specific and highly engaged community of self-trackers who seek to discover insights about themselves through their self-tracking practices.[62] To Neff and Nafus, the (lowercase) quantified self is a nebulous term that has been used in a variety of contexts as a catch-all term for any kind of Fitbit- or Apple Watch–wearing consumer. The Quantified Self community, however, has well-defined norms around the datafication of their bodies and health, particularly for self-discovery rather than just self-improvement. The Quantified Self community rejects much of the normative tendencies of some self-tracking systems and technologies as it refocuses attention on *knowing* the self rather than *improving* the self.

Scholars of the quantified self agree that the term is fundamentally associated with the notion of "self-tracking" (Lupton, 2016; Neff and Nafus, 2016). Self-tracking refers to the ways that people knowingly and purposefully collect information about themselves to analyze and reexamine. Media accounting, as I will show, is fundamentally a self-tracking process. To track means to trace the course or movements of something over time. Rather than using footprints on the ground, the qualified self uses media to keep track of the various events in our lives so that we can retrace who we are and where we came from.

Self-tracking scholars acknowledge the variety of ways through which people can track and have tracked themselves over many centuries. Neff and Nafus argue, however, that the contemporary focus on self-tracking intersects with two key transformations that significantly change the nature of self-tracking. First, technological advancements in mobile computing and sensors have enabled a variety of new kinds of information to be monitored, networked, and analyzed. Second, the biomedicalization of culture increasingly seeks biological or physiological explanations for life experiences. Together, these developments have given rise to an increase in self-tracking practices, particularly within the health domain.

The self-tracking associated with both the quantified self and the Quantified Self community describes turning bodies, experiences, or behaviors into data.[63] Gitelman argues that this mode of knowing emerged in the nineteenth century.[64] She demonstrates how documentation became the primary mode through which knowledge was made and circulated. Here, knowing became showing through documentation, which more recently we call data.

Overall, the difference between the qualified self and quantified self is a matter of degrees and focus rather than categorical difference. Both can be understood as self-tracking; however, the definition of self and the processes of datafication and mediation change the nature of how we engender and experience the traces we and others create.

From Datafication to Mediation

Although self-tracking is central to both the quantified and qualified self, a shift from the quantified self to the qualified self can best be described as a shift in prioritization from the processes of datafication to mediation. Viktor Mayer-Schönberger and Kenneth Kukier define datafication as the process of transforming a phenomenon into a quantifiable format so that it can be analyzed.[65] Recording and analysis are central to the datafication of self-tracking and are not unique to digital culture. As Neff and Nafus argue, quantification is not the defining characteristic of self-tracking—datafication is.[66] The transformation of datafication involves collecting, recording, and analyzing information that is self-tracked.[67]

The qualified self might not feel or immediately look like data, but it can be. A diary in the hand of the diarist might not feel like data, especially if they never read what they have written, but to the diarist's grandchild or to a historian it is. The diary or scrapbook as a collection of media traces can constitute data about one's ancestry or about a historical time period. It can be combined with other traces such as photo albums or archives to help us see patterns and personal as well as cultural meanings.

Whereas the qualified self can be understood as datafication, the qualified self can be better understood as mediation. Roger Silverstone defines the analysis of mediation as understanding "how the processes of mediated communication shape both society and culture, as well as the relationships that participants, both individual and institutional, have to their environment and each other."[68] On an individual level, mediation processes ask what it means to create media traces of one's life and the world around us, what the communicative functions or uses of the creation and circulation of such traces are, and, in turn, how such media traces affect us and our social relations. On an institutional level, mediation examines the political economy of the institutions which both enable and entice media accounting. Here, we can examine the roles of platforms and media technologies. For Silverstone, technology is essential for mediation.

> Media technologies are doubly articulated into the social both as technologies whose symbolic and functional characteristics claim a place in both institutional and individual practice, but also as media, conveying through the whole range of their communication the values, rules, and rhetorics of their centrality for the conduct of the quotidian.[69]

Within a mediation framework, the qualified self can be understood as both as a mediated reflection and refraction of us and the world around us.[70] Mediation presumes a bidirectional influence of media technologies onto us and us onto media. The qualified self is an assemblage of our media traces, enabling both multiplicity as well as contradictions to enter into our interpretations of selfhood.[71]

A shift from datafication to mediation can also suggest a shift from intrapersonal communication toward interpersonal communication. The intrapersonal communication of the quantified self is not to say that self-trackers don't get together and share their data or their practices; rather, the priority of their self-tracking can be understood as an intrapersonal communicative process whereby people self-track to understand themselves. The qualified self suggests a shift in communication prioritization, that is, the interpersonal communication of media accounting privileges the exchange, the audience, and the social relations of the qualified self. A flower seen on a walk and shared on Instagram or a family gathering photographed for the family album presumes mediated interpersonal exchange. Mediation, rather than datafication, reveals the ways that others feature prominently in our representations of ourselves. Similarly, audience and sharing are default to the qualified self rather than the exception. The social content and practice of mediation are central to understanding the qualified self as distinct from the quantified self.

A prioritization of mediation suggests that the experiences, activities, and events of others can become fodder for our own media accounting. The quantified self is often situated within an individual context, but the qualified self is often situated within a relational context. The qualified self is a qualified version of the self based on the qualities and qualifications of the media traces created by oneself and others. While the quantified self is more self-focused, the qualified self is focused more on others, both in audience and in content.

Dialectics of Media Accounting

In this book, I use dialectics to understand contrasting forces that continually push and pull our motivations, actions, and understandings of media accounting. Dialectical frameworks are common in both interpersonal communication as well as in media studies,[72] and help us explore the unresolvable tensions inherent in various processes. There are four dialectical pairs central to my definition of media accounting: public and private, individual and collective, work and leisure, and ephemerality and permanence.

Public–Private Tensions

First is the notion of public versus private. The contemporary moment is rife with debate about privacy and social media. What is privacy in these contexts? How is it manipulated by governments and the platforms themselves? In the smartphone era, when much of our communication occurs through mediated networks, how do we understand what is private or public? Rather than thinking of them as distinct spheres, a dialectical framework suggests they are oppositional forces that continually influence and shape our communication, our media accounting. The publicness and privateness are aspects of our media accounting that fluctuate and shift over time and depending on context.

As feminist scholars argue, contemporary divisions between public and private spheres as well as their associated gendered roles and responsibilities were never as clear-cut as we often assume.[73] Although we often associate women with the domestic or private sphere and men with the public sphere, Susan Miller argues that these distinctions overly simplify complex social relations and roles that characterized much of nineteenth-century American white middle-class life.[74] Women and men have long had to negotiate the continuum of publicness and privateness among our religious, political, and social interactions. Social media have not brought about the blurring of public and private life but have merely brought greater attention to it.

Individual versus collective is the second dialectical pair. Media accounting highlights the ways in which we must continually manage ourselves and others through the representations we create. We create media traces of ourselves for others. We create traces of others for ourselves. We read the traces of others to understand them and ourselves. Distinct categories of the individual or the collective do not convey the sociality at work in much of media accounting. There is continual social blurring that feminist scholarship has highlighted.[75]

Work versus leisure is the third dialectical pair. Again drawing on feminist scholarship,[76] this seeming dichotomy highlights the blurred distinction between what we consider work and what we consider leisure. Although industrialization of the nineteenth century brought about new distinctions between work and leisure,[77] women have never experienced such clear distinctions.[78] Media accounting can be enjoyable as well as taxing. The value of engaging in media accounting is both personal and collective, further blurring the distinctions between work and leisure.

Finally, ephemerality and permanence is the fourth dialectical pair. Media accounting is a process of taking life experiences and creating media traces of them. While the media traces are often static, their meanings are not. This dialectic is heightened in the contemporary social media environment, where so many of our media traces are digitally presented through our mobile devices. The seeming immateriality of these traces enhances a sense of ephemerality. The phone, however, is quite material, as are the networks and servers storing such data. Here, materiality and immateriality are replaced by visibility and invisibility. The more visible something is, the more permanent it feels. Media accounting allows us to manage the tensions that arise between the fleeting nature of the lived experience and the desire to hold on to such life experiences.

The Book

The book explores four practices associated with media accounting. By organizing it around practice rather than accounts, accounting, and accountability, I can draw out the ordinary uses, technological affordances, and historical parallels across media; identify the account, accounting, and accountability aspects within each practice; and focus on what people *do* with media. Each chapter explores a different practice; however, all practices are mutually enabled and constituted. Therefore, although certain social media platforms are used to explicate particular practices, any case could be used to identify each practice. For example, Instagram could easily be used in place of Facebook to explore any of the four practices.

Chapter 2 focuses on the first of four media accounting practices: the ways that people document and share everyday aspects of their lives through media accounting. This chapter draws on theories of ritual, routine, and presence to reveal how sharing mundane events and activities can be meaningful and important. I first review historical diarying to examine how people have used diaries for quotidian chronicling and sharing. I then draw on several key examples from my research on Twitter and mobile social networks to talk about how people share small bits of their everyday lives in ways that create meaning for them and those connected to them. Sharing what we read online is an important way to sift through and filter vast amounts of media. This kind of curation[79] allows us to share our own media consumption and is also an important part of performing identities (which will be discussed in chapter 3). Sharing the mundane also provides a new lens for understanding narcissistic critiques of social media. Therefore, I also explore issues of videologs, or vlogs, often considered narcissistic yet mundane, to explain not only sharing the mundane but our collective interest in consuming and bearing witness to quotidian media accounting. People have shared their activities, routines, and locations with others through media as means of social interaction and integration. Within a longer history of media accounting, we can begin to understand the motivations for sharing and reading social and location-based personal information. Media accounting provides a new lens through which to view the debates on narcissism and social media by putting the practices into a longer historical trajectory of meaning sharing.

Chapter 3 explores the role of media accounting in identity performance and work by highlighting how media have always been important outlets for identity expression. I integrate dramaturgical theory with notions of middle-class cultural identity and visual media accounting. I suggest that visual modes of identity representation are means of social interaction. Beginning with a review of the historical role of snapshot photography,

I show the early interconnections between media, the family, and identity. I explore the representations of identities through the creation and sharing of images, comparing early Kodak with Instagram. I then review the rise of consumer culture and scrapbooks at the turn of the twentieth century, discussing the importance of performance, consumption, and identity on Pinterest. Two important aspects of identity representations are explored. First, I argue that identity is not an individualistic cognition or state, but fundamentally a dynamic and socially enacted process revealed through media accounting. Second, I argue that the ways in which people make choices about the small scraps, snapshots, and posts of their media accounting reflect identity work. Particularly related to the family, I argue the identity work of media accounting is a form of invisible labor often taken on by women.

Chapter 4 explores the practice of remembrancing. Media traces have long been a tool for remembering activities and experiences. This chapter examines the various ways that people create media traces as what José van Dijck calls mediated memories.[80] Historically, travel journals were one of the most common types of diaries. They gave travelers a way to record new events and experiences to savor at a later time and share with others. Similarly, modern travel blogs and social media posts allow travelers both to share with others and to relive these traces after returning home. Remembrancing is also a way to create media traces of especially important events in our lives. This chapter explores the role of memorial photography and, in particular, postmortem infant photography as a means of understanding how and why we create traces of difficult experiences in our lives. Remembrancing is a media practice that ritualistically reinforces our social collectives.

Chapter 5 explores the practice of reckoning. Reckoning is the process of engaging with media traces to better understand ourselves and the world around us. This chapter examines the evidentiary nature of media traces. Drawing on Derrida's notion of the trace,[81] I examine the ways in which media accounting allows us to both prove and improve ourselves. I discuss how various media traces are used as evidence. I draw on the GoPro camera and its community of YouTube users to demonstrate how everyday people create and share their videos to document and prove that something happened. But I also argue that reckoning comes from the aggregated nature of media accounting. We can see patterns in our traces over time that we cannot glean from our lived experience. This chapter examines various tensions that arise when our media traces do not align with our sense of self, and describes a reconciliation process that we engage in through media accounting.

Finally, chapter 6 reviews the practices and dialectics of media accounting to explore what's really new about new media accounting practices. One of the benefits of placing social media into a media accounting framework is that it allows us to see similarities in practices across time and technology. However, it also provides insights into the key differences. In this chapter, I argue that the speed, size, and ownership of mobile and social media platforms are significantly different from previous forms of media accounting. I discuss the implications of these differences both for individuals and for our culture more broadly. I also discuss what I see as a postdigital turn in media accounting whereby the media traces we create are made analogue in a tactical way, as a means of regaining the power and influence of our media accounting to counter the current commodification of media accounting.

Taken together, the chapters of this book reveal long-standing communication and media processes. For hundreds of years, we have used media to talk about ourselves and about the world around us. We do this to connect with others, to fulfill social roles and responsibilities, to help us hold on to and commemorate the people and things that are important to us, and to better understand our place in the world. Mobile and social media help us do this today as our qualified selves are shaped and reshaped through our media traces and how we share them. In a very ordinary way, we have found great meaning and connection in using media to share our everyday activities and experiences.

2 Sharing the Everyday

One of the primary critiques about social media use is that people are sharing mundane and meaningless information. In one cartoon from the *Times Picayune*, a young man is depicted looking at his mobile phone saying, "It's a tweet from Amanda…She had cornflakes for breakfast and now she's putting her bowl in the dishwasher." His parents are sitting on the couch reading the newspaper as the dad responds, "Man, I hope newspapers survive…"[1]

This social critique is twofold. Part of the critique assumes that the son in the cartoon believes Amanda's tweet is news, or at least important information. Within a broadcast media paradigm, which includes printed newspapers and thirty-minute nightly news segments, this thought process is problematic because we assume media are finite. Therefore, the inclusion of Amanda's tweet would mean an exclusion of something more newsworthy. But the digital networked environment fundamentally changes this. These media are not finite—we can share and link to as many tweets and stories as we want. Amanda's tweet does not necessarily beat out serious forms of news in terms of space or time but it does in terms of attention. Depending on how intimate they are, I might suggest that anything "Amanda" says or writes regardless of the media platform would take this young man's attention away from "serious news."

Another aspect of this critique is that it suggests that Amanda believes her sharing what she had for breakfast and how she cleans it up is actually newsworthy. Why otherwise would she write about the details of her everyday life and share it with others publicly? Within a broadcast paradigm, media are used for professional news, entertainment, and persuasion, mostly in the forms of advertising. We can use the transmission model of communication to explain these three types of broadcast media—in each type, a message is transmitted from a sender to one or many receivers. However, social media and media accounting do not follow this framework. If we think of sharing the mundane within Carey's ritual model of communication then these communicative acts represent a rich and complex form of social interaction and integration.[2] This suggests there are many reasons beyond news, entertainment, and persuasion for why people document and share the quotidian aspects of their lives. Indeed, sharing the everyday is first a media accounting practice. By examining how and why people have chronicled and shared the quotidian before social media existed, we can better understand why people do it today.

Historical Aspects of Documenting and Sharing the Quotidian

Diaries are probably the most common format for documenting everyday life in the past two hundred years. Fothergill identifies four genres of English diary writing—journals of travel, "public" journals, journals of consciousness, and journal of personal memoranda[3]—all of which are forms of media accounting.

Journals that chronicle travel have been particularly common because the journey invites journaling. Fothergill notes, "The travel journal appears to have been one of the earliest types to achieve the status of a recognized form in which to render one's experience."[4] Travel accounts also have historically served as an important source of information for an audience at home so that they could learn about the world beyond their immediate surroundings through the written and shared experiences of others.

What Fothergill refers to as "public" journals are regularly written entries typically associated with a task or profession, such as ships' logbooks or transaction logs. For women diarists, the kind of "public" diary that Fothergill describes has historically been associated with tasks, duties, and types of work that are often domestic and personal. Feminist scholars such as Margo Culley and Susan Miller show that women's diaries also served a social function whereby they chronicled not only the activities of the diarist herself but also the events and activities of the household or community.[5]

Fothergill's third category, journals of consciousness, focuses on the inner life of the writer. Journals of consciousness or religious diaries were common among Puritans and Quakers throughout the eighteenth century as a form of writing meant for self-development and spiritual self-discipline. Keeping a diary was a way to become a more pious individual. Writing about one's moral and spiritual self not only reinforced one's beliefs but also could actively shape future acts. We also see this aspect of diary keeping taken up at the end of the nineteenth century in Victorian households where parents would read the diaries of their children as a means of reinforcing proper behavior.[6]

Personal diaries have been defined as first-person writings about the activities of the author himself or herself written about the present in the present. This is contrasted with autobiographies or memoirs, which also involves the author writing about himself or herself, but these are often retrospective rather than presentist in their accounts: "The diarist is to the day as the autobiographer is to a life. What the text lacks in perspective, it gains in immediacy."[7] These diaries are about the daily activities, musings, and reflections on the events of the day. Personal diaries are differentiated from journals of consciousness in that they are grounded in everyday activities and events rather than the emotional or religious reflections and prescriptions for the future.

Feminist historians have suggested that diary keeping of ordinary women reveals the important but often overlooked roles and experiences of women.[8] Whether it be chronicling physical and emotional demands of childbearing or the small details of running the household, women's diaries give voice to their everyday activities. Diaries were social in both content and practice, as a way to maintain kin relations. Young women would keep diaries when they got married and send them to their parents. When friends or relatives would visit, it was not uncommon to read through one's diary together as a means of catching up on the events of the household. Therefore, by giving voice, media accounting was a dialogic rather than monologic form of communication for these women. The qualified selves represented in these diaries are personal as well as relational.

Genres of diaries of course blur. Religious, public, personal, and travel writings can all occur within one diary. Some people may engage in just religious or just travel personal writings. Or they may do both but at different points in their lives or in different notebooks. All of these diary writings are forms of media accounting regardless of the type and content.

Diaries and Media Accounting

One of the many reasons why the diary is a helpful comparison to other forms of media accounting is that it has historically been defined by the "entry." Unlike other kinds of texts, the diary is "a text segmented into units marked by date and day."[9] This kind of segmented text is also characteristic of media accounting more broadly. Though most social media platforms refer to the entry as a "post," the segmented nature of writing text or writing with pictures is a fundamental structure of media accounting.

But how does the diary "entry" differ from the social media "post"? Both are fragmented and characterized by their discontinuity, but the entry conveys a sense of recording. To make "an entry" means to enter information into a list or record; to post means to show or display a message. As such, an entry highlights the registered nature of information, whereas "the post" highlights the public nature. The post does not immediately convey the same registered or recorded nature of the message. While posts are indeed *entered into record,* that of the social media networked database, the term belies this aspect of the text. Similarly, the term "entry" does not convey the public nature that characterized many diary practices throughout the eighteenth and nineteenth centuries in the United States.

One of the key features of all diaries, regardless of genre, is its day-to-day nature, in terms of usage and content. The diary entry is an account of how one has spent his or her day. Occasionally the diarist will write a catch-up entry, which not only describes their experiences of the day but of yesterday as well, but there is a commitment to the present, however roughly defined. Therefore, it is argued that the diary constitutes a more accurate and authentic account of everyday life than other forms of life writing because it is not clouded by memory or filtered by time.

This is not to suggest that the diary or any form of media accounting is necessarily the truth or fact. Fothergill argues that diaries "are not truthful, but they are actual."[10] This subtle distinction conveys the nearness of the writing to the experiences of the diarist. The events and activities chronicled may not be the facts but they are actually what the diarist felt, thought, believed, or wanted to chronicle at that moment of writing. While there will always be details left out of the diary, it represents an *actual* account of someone in their own hand.

Therefore, the post and the entry both convey a presentism, which in turn reflects much of their value. While Fothergill notes that entries may be highly edited, they are not revised or rewritten in the same way that other texts are. The entry and the post are both written at particular moments. "What we are reading now we need to be assured must be what was written then."[11] That is, the texts are written at the time when they indicate they were written. The content also reflects this presentism—what happened that day, what happened that moment, or what happened just prior. The presentism of what is written about gives the text veracity and authenticity. Unlike the autobiographer who has time to reflect and then write about himself long after events of the day have passed, the diarist lacks time for reflection and merely reports on what has occurred.

A common aspect of diaries and media accounting is the frequent omission of the first-person pronoun. Rather than writing "I went to the store," often people write, "went to the store." While such an omission may be interpreted as just a shorthand, Fothergill suggests that some diarists are aware of the social norms against speaking incessantly about oneself and perform accordingly: "Perhaps the suppression by some diarists of the 'I' that would otherwise govern verb after verb, page after page, is a gesture of self-effacement, a tacit apology for the appearance of self-preoccupation."[12] I want to suggest that similar norms exist across media accounting more broadly. Today we see a lack of "I" in tweets and Facebook posts, which may represent this continued normative expectation.

Routine and Ritual

The dailiness of media accounting can often become part of an everyday routine. Get home, have dinner, put the kids to bed, go on Facebook or write in your diary before going to bed. This routine gives us a time to reflect and chronicle what happened that day or what's going to happen. The routinization of media accounting is important because by repeating this, it becomes second nature to us to the point where we automatically go through the motions. As such, a lot of assumptions and meanings about what it means to engage in a particular behavior become tacit. In taking for granted our media accounting, we obscure its nature and meaning. We found in a cross-cultural study of smartphone use that people would just "check their phones" in moments of pause or transit or waiting.[13] They reported often not realizing they were doing this—it was just a force of habit.

Media accounts are sometimes in a process of becoming routinized or ordinary. Sometimes there are deliberate efforts to begin media accounting. People will join a new social media platform because their friends told them about it, or they will start keeping a diary because someone gave them a nice journal as a gift. I know a number of faculty who began a blog for their sabbatical or students who blog during study abroad. Occasionally people will deliberately stop media accounting, but most often people stop media accounting because they just forget to do it.

I want to suggest that media accounting can play the same role as other everyday rituals and become a site for transformation. Catherine Bell argues one of the important aspects of ritual practice is the dialectic between thought and action.[14] Rather than suggesting that the two are binaries or dichotomous, practice allows us to see them as mutually constituted. Media accounting becomes the site for integration, where thought and action come to bear. It enables us to document thoughts and actions for ourselves and others, then to further reflect or take action to change or bring them to life.

Media accounting facilitates the syntheses and integration of conflicting forces. Bell writes: "Ritual is a type of critical juncture wherein some pair of opposing social or cultural forces comes together."[15] Media accounting facilitates the integration of thought and action but also moments and memories, and the lived and documented. Media accounting becomes a ritualized means of integrating these forces toward the continued development and shaping of the qualified self.

Media Ritual

Much of the research focusing on media and ritual has primarily looked at mass media. It has been explored both in the ritualized ways people consume television and through its production.[16] Dayan and Katz's concept of media events are probably the most well-known integration of media and ritual, as they are out of the ordinary and collectively consumed.[17] All regular programming stops and we all focus on the same spectacle like the Olympics or the wedding of Prince William and Kate Middleton. These moments of collective media consumption are transformative, as they become sources of social integration.

Karin Becker argues that media help to transform a public event into a ritualized event through its documentation. She argues that the documentation of an event through media helps to ensure that the "event carries symbolic significance beyond the bounded sphere in which it is unfolding." Like Becker, I suggest that media accounting can make experience and events into rituals: "Being documented is part of the play that marks the event as a ritual."[18]

Documentation, however, can transform mundane activities into media accounts. Becker's examination of media rituals and public events can be applied to personal or private events and activities. When people document birthdays, the first day of school, vacations and holidays, the ritual nature of the events become reinforced through media accounting. These are moments of everyday transformation and social integration. Media accounting actually plays a very important role in acknowledging, verifying, and validating the meaning of the everyday in the same way that Becker suggests media transfer public events:

> The passage character of media can be seen in the ways activities of documentation transform private and individual behaviors and experiences into public and collective ones. Through their framing activities, media shift both private and public action and behavior into the realm of performance.[19]

Becker suggests media have the power to transform public events into performative space. Even when people would use diaries to document the quotidian aspects of life, media accounting transforms these activities into ritualized performances of the mundane.

Media accounting can also transform the private space into a performative space as well. Media accounting transforms everyday activities into something that can be consumed at another point in time and potentially by others. Its ritualistic documentation is a form of cultural performance that is evidence of a broader social order. Therefore, media accounting is both the active maintenance of and the performance of social order. Carey makes the distinction between representations of and for reality. He argues that symbols *of* present reality, as opposed to symbols *for*, which "create the very reality they represent."[20] In this way, media accounting as a form of ritual communication can be both the representation *of* social order and the representation *for* social order. The representations of ourselves in media accounting reflect social norms but also shape social norms. Thus, the qualified self is simultaneously made up of representations of us and for us.

The social significance of everyday activities can be communicated through meta-narratives as well. Becker suggests that media experts and commentators help to further transform media coverage of an event into ritual by providing a layer of reflexivity about the event.[21] This kind of commentary can occur within media accounting in three ways. The first is by the author himself or herself. Unlike journalists who are not supposed to provide their subjective thoughts, feelings, and interpretations about activities they are covering, individual citizens are. In fact, their perspectives and opinions are expected to help other consumers of media understand and interpret their media traces. Part of why someone would blog about an event is to share his or her version of it. In our Twitter research, we found that most often people would not only report on new events or activities but they would also comment or reflect on those activities.[22] Thus people not only provide documentation of an event or activity but also they provide a meta-narrative to the activity, such as when people live tweet an awards show. In this case, one event can have many accounts of it, which when aggregated can reveal broader insights beyond the official (meta-)narrative.

The second way that we see media transforming events is through the metacommunication about media accounting itself. Sometimes people reflect on what it means to blog, post, or tweet. What do we think about what we're doing? Why are we posting? What do we enjoy about keeping a diary? What do we get out of it? What does it do for us? When people reflect on these kinds of activities, this metanarrative further transforms the lived experience into a ritualized account.

A third way that media accounting allows for reflexivity is through the commentary of others. Media accounting is not monologic; rather, it exists within a larger social dialogue. Sometimes this is other people leaving comments to a Facebook post or replying to a tweet. Historically, people would write in the margins of one another's diaries.[23] We also dialogue face-to-face about the media traces we create. When we look through photo albums or scrapbooks together, we talk about them. While often thought of as monologic in character, media accounting engages in dialogic and reflective metanarratives, which further reinforce the transformative nature of the ritualized performances.

The social or group aspects of media accounting cannot be forgotten. While it may seem that individuals are transformed, ritualistic media accounting is part of the social process of integration. Hillis reminds us that ritual theory is a helpful framework for understanding online life more broadly.[24] While he argues that networked media give rise to new forms of media rituals, sometimes these are anti-hegemonic in nature, anti-rituals or fetishes. Nevertheless, there are still normative social centers, and media rituals like media accounting integrate toward them.

Cases

Let's turn to two contemporary examples of media accounting that highlight the ways people ritualistically document and share the everyday.

One of our original studies examining media accounting was a comparison of Twitter with historical diaries.[25] Drawing on Fothergill's analyses of historical diaries, we systematically coded a random sample of tweets to understand who was being discussed, what was being discussed, and how was it discussed. We were able to draw a random sample because, at the time of the study in 2008, Twitter was not quite the "firehose" of tweets it became eight years later. In order to make the comparison between past and present, we developed categories and codes for how we knew people had used diaries historically according to Fothergill and Culley in order to apply them to tweets.[26]

For example, some religious or personal diaries primarily focused on the diarist himself or herself, but other more public or travel diaries were much more social in nature. In our Twitter analysis, we coded tweets as being about the author or other people. We found that about two-thirds of our sample were about the author him or herself, but that over 40 percent of those tweets also involved another person or persons. This suggests that people are not just talking about themselves but invoking others when they do.

Because historical diaries often chronicled the quotidian events we coded tweets for common historical diary topics: food/beverage, health, weather, sleep, family, religion, as well as activities. Where possible we subcoded activities into home, work, and outside of home and work. Since the late nineteenth century, diarists were often encouraged to keep track of their novel reading in their diaries,[27] so we also coded for media. Altogether, almost 70 percent of our sample included these historical diary topics, of which activities (41 percent) and media (35 percent) were the most frequent.

Last, we coded for the narrative style in tweets. Early nineteenth-century secular diaries in the US tended to focus on the events of the day and present information in a truncated and explicit way, characterized as an accounting style. While late nineteenth-century diaries tended to focus on thoughts about the day's events and include more introspective and reflective kinds of entries. We found that 75 percent of our sample involved a commentary or reflection and that 62 percent of the sample reported on or shared recent information about the subject of the tweet in an accounting diary style. Between the reflection and accounting styles, we were able to include over 95 percent of the sample into one or both of these categories. This suggests that tweets in our sample seem to have both accounting and reflective elements.

Overall, this study suggests that people on Twitter write and share things about themselves, activities that they have just done or are about to do. We see that people use Twitter to document and share things they just read or watched. And like diaries, they aren't just tweeting about themselves. They are writing *about* others as they write about events of the day.

Interestingly, one of the items that we were unable to code reliably was news. This was something that people had written about in historical diaries that we wanted to replicate in our study. However, what is considered news is highly contextual. Our sample was created before including @username within a tweet caught on as a way of tagging people in tweets. Therefore, there were numerous examples in our sample of people speaking to a general Twitter community, or "twitterverse," in ways that made it difficult to assess the "newsiness" of a tweet.

Like diaries, part of the value of Twitter is that it chronicles events and activities in "real time." The dailiness of diaries is now accelerated to the moment on Twitter. Though this may seem like a big difference, the more significant difference between Twitter and diaries within a media accounting framework is when and how they are consumed. Even though both Twitter and diaries are written about the present, only Twitter is also consumed or read in the same present as well. This allows for quicker responses, which can then inform and feed back into the chronicling of the activity or event itself.

We also supplemented the content analysis with an interview study in which we asked people about how and why they read Twitter. One the key reasons people read their feed was to see what's going on among their friends and celebrities. It was expected that both friends and famous people would be sharing what was happening and what they were doing. Sometimes people in our sample would respond, but often they found value in just knowing what was going on. This was an example of intimacy emerging from the sharing of small bits of information about everyday experiences and activities. Users could become part of a larger ritualistic process of communication reinforcing social relations.

While Twitter is often the social media poster child for chronicling the mundane, I want to turn now to another case that also reveals the ways that the quotidian aspects of media accounting transform the everyday into a media ritual.

ThePointlessBlog

One of the many popular video blogs (vlogs) on YouTube is by Alfie Deyes, a young British man who began vlogging in 2009 at the age of fifteen. In 2016, ThePointlessBlog had over 5.4 million subscribers and over 436 million views. But like many vlogs, it had a humble beginning.

Alfie began his first ever vlog post with "This is England and this is me in England." From the very beginning, Alfie seemed to be talking to an American audience, and so the articulation of his country was very explicit. Throughout his early posts, Alfie implicitly and explicitly addresses an American audience by explaining things that are English.

Alfie's YouTube channels are helpful to analyze for several reasons. First, he was a self-proclaimed "ordinary" young man who started a video blog because he enjoyed watching them himself. He began by documenting and sharing the small and everyday aspects of his life. Second, I want to suggest that the value of quotidian chronicling is not just something that "ordinary" people do, but something that everyone does within this genre. The small details of famous people or people who are professional media accountants still rely on the same notions and commitment to the quotidian in ensuring the value and authenticity of their media accounts.

Like many vlogs, much of Alfie's vlogging can be characterized as conversational, where Alfie talks directly to the viewer. He does this most often from somewhere in his home. This placement of where the videos occur is actually quite important to the videos' sense of intimacy. Sometimes he is in his kitchen or sitting on his sofa, and sometimes he's sitting on the edge of his bed or at his desk. These are all his places, private places. Occasionally, he'll do a video lying in bed with his shirt off and say goodnight right before he turns off the light. The personal places from which he vlogs contributes to the sense of intimacy with his fans.

The regularity of his posts is also very important. This became explicit in 2010 when he ordered a new camera that was late in delivery and so he missed posting one week. He made a one-minute and twenty-five second video explaining why he hadn't uploaded and apologized, explaining how it is not an indication of his lack of interest. Like many other vloggers, he posts weekly and has started other YouTube channels (a daily vlog and a gaming vlog), which have over a million subscribers. Weekly, not daily, posts tend to be fairly normal for vloggers. This is likely because the amount of time it takes to edit a video is much longer than the amount of time it takes to edit text. Fothergill describes a diary as edited but not revised, in that the content was carefully selected during composition but not changed at a later point in time. The regularity of media accounting, whether it be daily or weekly, as well as editing, help to reinforce its ritualistic nature. Both aspects reinforce the transformation between thought and action, between presentation and representation.

Compared to his main vlog, Alfie's daily vlog is very similar to a daily diary in that he is capturing what happened that day. In his own words, "I'm gonna be uploading some ... well, I'm not really sure yet. Maybe some little vlogs or something on littler topics that can't really go on my main channel because they're not big enough. Maybe some little personal videos, or just talking to you guys." His daily vlogs are shorter and more mundane than this main vlog.

Visually, the daily vlogs are also distinct from the videos on Alfie's main YouTube channel. They are not as highly edited or scripted, and may have lighting and sound levels that fluctuate. The videos are often handheld shots of Alfie by Alfie, a kind of video selfie, right before something is about to happen or right after something has happened. His daily vlogs are mostly him recounting his activities, meetings, and social engagements. In addition to his house, his daily vlogs are also shot in transit, either in the car or on the train. As Amparo Lasen argues, the visual depictions of mobility are central to understanding the affective contemporary experience, where mobility and emotional availability are closely linked.[28]

Sometimes Alfie's daily vlog includes the behind-the-scenes outtakes of his main channel videos. This results in two aspects for the viewer. First, it shows the amount of work that it takes to make videos for his main channel, encouraging the viewer to watch that channel. Second, it reveals the backstage to his main vlog, thus providing another avenue of intimacy into Alfie's life. This backstage video is not as scripted and thus provides a more authentic if boring sense of Alfie. The details in his daily vlog reveal his process, the details that go into making his main channel's video.

This intimacy is also reflected in the ways he addresses his audience within the video itself. For example, in 2010, he ended his blogs by asking viewers who liked the video to click on the like or thumbs-up button, but by 2015 he doesn't have to. Instead he ends his videos with: "Thanks so much for watching. I love you."

Vlogging, like other forms of media accounting, can be misunderstood as documenting every aspect of someone's life. But this is never the case. There are always things that aren't chronicled and documented. But both the regularity and the quotidian details of media accounting can suggest a comprehensiveness even if shorthand.

Jean Burgess and Josh Green suggest that YouTube is really made up of two YouTubes.[29] One YouTube consists of traditional media companies using YouTube as just another channel through which to reach their consumers, whereas the other YouTube consists of user-generated videos and content. Of course, they recognize that these categories are problematic and not clearly differentiated, especially on a site like YouTube, where large companies partner with vloggers to leverage user-generated content. Nevertheless, they argue, these are helpful analytical distinctions. They argue that vlogs are exceptionally important on YouTube:

> Vlogging itself is not necessarily new or unique to YouTube, but it is an emblematic form of YouTube participation. The form has antecedents in webcam culture, personal blogging, and the more widespread "confessional culture" that characterized television talk shows and reality television focused on the observation of everyday life.[30]

Of course, vlogging can be put into a much longer trajectory of personal chronicling for a public audience. The publication of diaries throughout the seventeenth and eighteenth centuries in England was widespread; therefore Steinitz argues that many diarists historically wrote to a public audience.[31] Some diaries became widely read and popular, such as Samuel Pepys's, but others were not widely read. Similarly, some YouTubers like Alfie are able to become wildly popular, but others will only ever be seen by a few.

Part of the allure of vlogs and diaries is that they provide a lens into someone's life at an angle and from a perspective that we don't get in our lived experiences. When we read or watch the mundane aspects of the lives of others, it helps us to think about our own lives as well as to escape them.

The Pointless Book

In August 2014, Alfie announced a book project called *The Pointless Book*. In his video, he describes the book:

> Something I love to do is store memories, I love storing history, things that I've experienced so that I can look back on them later. So I want to create something for you guys to look back on. Something that's going to be fun, something that is going to be useful. And you can all do it together. Not vlogging. Something else.[32]

He encourages people not to simply read the book, but to interact with it. Readers should not start at the beginning and linearly read it but are encouraged to flip through and see what excites them. "This is like a journal for a year of your life or a couple months of your life. And you can just fill it out with your friends."

Throughout the book there are activities for readers to engage in with others. For example, one page instructs readers to write something on it and swap it with a friend. The page is perforated to further facilitate and encourage this exchange.

> As you know, interacting with you guys is one of my favorite things to do, so I have curated a book, a kind of journal thing, for you guys to complete and have fun with.... It's full of challenges, and games to do some of which I've done in my videos, pages to fill out. Video diary kind of things.[33]

Like many other YouTubers, Alfie's book is an extension of his vlogs, another way to connect with his fans. There is no Kindle version of *The Pointless Book*; it is available only in paperback. This book is both a place to keep memories and a place to create memories.

The Pointless Book is particularly interesting because of its materiality. It is meant to be held on to—in that it is meant to be both held in one's hand while writing, and kept over time. It is directive in that it provides people with prompts; for example, its dream journal page encourages people to write what they dream about.

But the language around pointlessness of both the blog and the vlog matters and warrants closer examination. First, it suggests that Alfie doesn't take himself too seriously. Much like Alfie's blog, the book is meant to be fun and entertaining. It's not too serious or worried about being productive or meaningful. It's supposed to be fun and easy. For the primary watcher of Alfie's vlogs, the young adult, books are often associated with their work at school. He is offering a different kind of book.

Second, "pointless" deescalates Alfie's promise to his audience. To say his musings in this video or book are pointless lowers peoples' barriers to persuasion. He's not trying to teach us or convince us. He's just here to hang out with us. The pointlessness of the everyday is not threatening but comforting.

The quotidian therefore becomes an aesthetic, a value, a practice that is a primary characteristic of media accounting. The "pointless" is about transmission, not about the ritual of coming together and sharing the everyday. Documenting the pointless things is a form of sharing intimacy, authenticity, and honesty, which connects the human spirit.

Tension

One of the key concerns about media accounting is its narcissistic value. This is especially true for vloggers—it seems that to create a video blog of yourself you must be self-obsessed. Maggie Griffith and Zizi Papacharissi argue that "personal vloggers generally talk about themselves, what interests or concerns them in their videos. This inward focus certainly satisfies a narcissistic desire."[34] Here, narcissistic is defined as being overtly focused on oneself and in need of attention from others. But Griffith and Papacharissi acknowledge narcissistic nature of vlogs is about generating some sort of audience, which is fundamentally about making a human connection.

Sherry Turkle has also argued that mobile and social media represents a concerning narcissistic turn.[35] We now must document our lives through media before we can understand ourselves whereas before the rise of social media we could just live without the need to document. Turkle suggests that before social media we could understand ourselves based on our lived experience rather than the narcissistic need to share our lives with others. So, what makes vlogging narcissistic?

To keep a diary and chronicle what happens in one's life does not seem to raise the same narcissistic concerns that vlogging does. People seem to not have problems with writing about or photographing the minutiae of one's life. At worst, one could be called obsessive-compulsive, but not narcissistic. Therefore, it must be the sharing of personal details that warrants condemnation. To assume that others would be interested in the details of one's life is often where vlogging or any form of media accounting seems to raise pathological concerns. Webcam culture also has drawn similar scrutiny.[36] To document one's life is fine, but to make one's life public is what seems to make it "bad."

Some have argued that a vlogger who appeals to a large audience cannot be self-obsessed. Kate Murphy of the *New York Times* argued in 2013 that to be successful as a vlogger, you cannot be narcissistic. "If you're an aspiring video blogger, remember, it's not about you, it's about who is watching you. Be conscious and considerate of your audience and its needs, rather than getting mired in your own egotism or insecurity. (It's good advice for life but essential to making quality video.)"[37] And indeed much of the "how-to" literature around creating YouTube videos (and other guides for creating "good" social media) is about sharing information that will be of interest or value to viewers and followers.

So, is media accounting necessarily narcissistic? If we define narcissistic as the *Oxford English Dictionary* does, as excessive self-admiration, then no. Indeed, much of media accounting has been about self-scrutiny rather than admiration. But if we define narcissism as Christopher Lasch does when he wrote about the narcissistic culture in America, long before the rise of social media, then it might.[38] Here Lasch argues that the narcissistic culture is obsessed with the present rather than posterity or the past. It also is characterized by a therapeutic sensibility, according to Lasch, which replaces organized religion and focuses on the self rather than community. However, like Papacharissi argues,[39] I want to suggest that, while media accounting is a form of self-expression, that does not necessarily mean it is pathological, as Lasch would suggest. Vlogging, and social media more broadly as a form of media accounting, does not necessarily represent a narcissistic turn in our society. We have often written about ourselves. Indeed, what else should we write about if not what we are the sole expert on? Might it not be more "pathological" to indeed write about and share things that we do not know much about? At least when people share their own experiences, they know about it.

Media accounting's focus on self-centeredness and introspection does not necessarily reveal pathologies. Papacharissi argues social media is situated within a narcissistic cultural context, but social media does not therefore have to be bad:

> Narcissism here is employed to understand the introspection and self-absorption that takes place in blogs and similar spaces, and to place these tendencies in historical context. Lasch's work, over psychological research on narcissism as a personality disorder, serves an apt starting point. Narcissism is defined as a preoccupation with the self that is self-directed, but not selfishly motivated. Narcissism is referenced as the cultural context within which blogs are situated, and not as a unilateral label characterizing all blogs.[40]

Like Papacharissi, I argue that the self-centeredness of social media, and media accounting more broadly, is common characteristic of a set of practices. Rather than pathologize media accounting as narcissistic, we can understand the documenting and sharing of quotidian experiences as an important way that we situate and connect ourselves to the world around us. Documenting our lived experiences and sharing them with others transforms our experiences of the everyday. As we chronicle quotidian events and activities, we engage in a social process. As we document everyday activities, media accounting transforms them into ritualized performances. This is easy to see in the case of Alfie's blog but is just as true for any media accounting tweet. The ritual of media accounting is a transformative one that relies not on the transmission of information but in the social interaction and integration that occurs through the production *and* consumption of media accounting.

Why Do People Read and Watch Media Accountings?

People love to read the intimate and everyday details of others' lives. We certainly want to see the posts, entries, photos that include us. But people will also read about what celebrities had for breakfast or how historical figures traveled across counties and countries, just as they will read the diaries of past relatives who document the weather and their own travels. The reading of media accounts is not new, but has occurred for hundreds of years.

There are many reasons why people would value or enjoy reading the media accountings of others. People who are related to the author in the some way, such as distant family members or friends find that reading the mundane life details is a way of reinforcing kin connections. The sharing of quotidian details is sometimes not about the content itself but about the act of sharing. This is often described in interpersonal communication as phatic communication, or communication whose content is less important than the communicative act itself. Calling "just to say hi" is a kind of phatic communication where the act of calling reinforces the relationship more than the content of the message. A grandson reads his grandfather's diary and gains a sense of intimacy with his grandfather and his family more broadly. Knowing the daily routines and events of someone's life can build intimacy between people.

People who have ostensibly no connection also can value the reading and watching of media accountings of strangers because it can be both entertaining as well as humanizing. We can see ourselves in the quotidian aspects of others' lives. Sometimes we can see connections to other groups of people from whom we had felt ostracized. For example, on May 31, 2016, National Public Radio aired a story from Radio Diaries, a nonprofit that helps to create podcasts of people's everyday experiences, about a Muslim girl. It was titled "Diary of a Saudi Girl: Karate Lover, Science Nerd ... Bride?"[41] The story is an account of Majd Abdulghani going to an Outback Steakhouse restaurant with her brother. She explains that she ordered salmon and her brother ordered chicken-fried chicken. She describes the sections of the restaurant, one for men only and one for women or mixed genders, as is typical in Saudi culture. The story includes an exchange between her and her brother that both reflects on his role as *wali*, the Arabic term for the masculine figure within the family who is in charge, as well as his role as a teasing older brother. It's through the chicken-fried chicken and the brotherly teasing that an American audience can see intimate connections with someone from seemingly such a different culture. The diary genre builds a sense of connection and intimacy through the sharing of the everyday.

Sometimes the reading of historical diaries can seem like a puzzle. Robert Fothergill argues that to the devoted reader, "it is precisely the coral-like aggregations of minimal deposits that become addictive. In propagating their passion, diary-enthusiasts tend to rhapsodize over the charm of long-lost trivia. A kind of cult develops around some of the more lovably artless chroniclers of their little lives."[42] Diaries can also be connected to other diaries. For example, the diaries and letters of Jane Franklin became part of Jill Lepore's historical analysis of Ben Franklin's sister.[43] The value for the reader or viewer of the snippets of other's lives comes in seeing its entirety. It is the aggregation of such writings within diaries as well as across diaries that can reveal patterns and broader insights into the individual chroniclers and their worlds. This piecing together of small aspects and non events can together communicate much more than their individual components.

Why the Quotidian Is Necessary for Epidemics and Revolution

It is important to both document and share the quotidian aspects of life for several reasons. First, activities, events, and behaviors that occur every day reveal important routines that are central to the social world. Much of what we understand is central to a culture occurs in these small and regular behaviors. While seemingly mundane, they can reveal much broader trends in social values. Second, in order for large or momentous events to be captured in the media accountings of people, they had to have been documenting regularly beforehand. Much of what we know about events and life in the eighteenth century is because of personal diaries. In order to capture big events, people had to have been documenting other activities and events—small events, everyday comings and goings of the household. Only then were people in a position to capture the momentous events of the time.

Much has been written about the role of mobile and social media in the 2009 Arab Spring revolution.[44] But little has been written about what happened before or after these momentous events. A media accounting framework spotlights the documenting and sharing of quotidian experiences that both precedes and follows momentous events. As such, the everyday chronicling of life through media accounting provides the backbone or underlying practice that allows for smart mobs or political organizing to occur and effect larger-scale change.

Summary

This chapter has outlined the media accounting practice of sharing the everyday. I situate contemporary vlogging and social media posts into a longer historical context about the ways people would use diaries to catalog and share occurrences in their lives. I draw on two notions of ritual theory to understand the practice of sharing the quotidian. First, I argue that this practice of media accounting is best understood through a communication ritual model rather than transmission model of communication, because the content is often less important than the intimacy that documenting and sharing such meaning represents. Next, I use ritual to understand media accounting as a meaning process, one that transforms thought into action and experiences into traces. Using Karin Becker's work on media rituals, I argue that the ways we use social media to comment on and share what we are reading, watching, and doing is a form of meta-narrative that further ritualizes media accounting and the qualified self.

Using ritual communication as a framework for sharing the everyday, I then describe two contemporary cases of media accounting that highlight this very practice. In our empirical content analysis of tweets, we found many people sharing information about recent activities and commenting on them. Similarly, the trope of Alfie's blog surrounding pointlessness serves to depoliticize and deescalate expectations regarding the informational and entertainment value of his videos. The quotidian aspect of his content and style led to a kind of intimacy between blogger and audience. This occurs not just on YouTube, but on all forms of social media as well as media accounting more broadly. The presentist accounts of ordinary and everyday experiences enhance their authenticity and value, making them interesting to read or watch because it highlights our communal nature.

The practice of sharing the everyday highlights several media accounting dialectics. Most predominantly, sharing the everyday blurs the public and private. While many of us today think of diaries and other forms of life writing as immensely personal and private activities, historical scholars reveal the exchange of such media accounting. The transformation of thought and action into media traces enhances their shareability. The meta-narratives and commentary about our media traces further transform them into a ritual communicative act, wherein the act of sharing such details reinforces our social connections. Indeed, the sharing and exchange of media traces raises questions about whether they can be considered public or private. What is personal is not always private and kept to oneself, but can be shared and known among others. The sharing of everyday details of life may be mundane but their intimacy cannot be overlooked. Placing social media into a longer historical context of intra- and interpersonal media use helps us to complicate notions of public and private that we see on social media today.

The chronicling and sharing of the everyday also blurs self and other. The content of diaries were not always just about the self, but often included the activities and goings-on of people in the diarist's life. This has been particularly true for women throughout history. Coupled with the intended audience of such accounting, sharing the everyday highlights how when we write or tweet about those around us, we blur notions of what is us and them—we reveal our social character. We write both for and about others in our media accounting in ways that blur our subjectivities. The qualified self has always been a relational self both in content and practice.

3 Performing Identity Work

The second media accounting practice is identity work. An important part of our communication with others is in performing our various social roles. For example, I am a professor, a wife, a daughter, a mother, and a friend. Therefore, how I act will change depending on whom I am with. In dramaturgical theory terms, our performances change depending on both the role we are playing and the audience we are addressing.[1] How I act around my students may be different from how I act around my kids, which may be different from how I act around my friends. All of these performances are sincere versions of me, but just different aspects of who I am and the various social roles and identities I perform. Much of how and why we communicate can be understood through an identity performance lens.

Performing identities is an important aspect of media accounting because it helps to explain why people create the kinds of media traces they do. They post photos of the people in their lives because it is part of their relational identity performances. Sharing content of various forms may be part of what it means to be a foodie or a liberal. The post reflects the identity of the person who posts it. A dramaturgical perspective of communication suggests that there are interaction rituals that people engage in through their communication.[2]

For example, on Father's Day, I posted a photo of my dad on Facebook wishing him a happy Father's Day. Similarly, I posted a picture of my son on his birthday wishing him a happy birthday. While it may seem that I am just communicating with my dad and son through Facebook, neither my son nor my dad are on Facebook. Neither of them have ever had a Facebook account (although I suspect when my son gets older, he might). The meaning of such messages cannot be understood through a transmission model of communication (e.g., sender → message → receiver) because the targets of such messages are not there to receive them. Nevertheless, my presumed audience of friends, colleagues, other family members are in mind as I post. Therefore, such posts must be understood through Carey's ritual model of communication.[3] A ritual model presumes that the purpose of communication is to reinforce the social order and the "maintenance of society over time."[4] Posting such messages is part of the performance of my roles as a daughter and mother that reinforces the social structure of the family. The roles of mother and daughter are not mine to make up, but to enact and model based on my own experiences of others' performances within the social world.

Representations of Identity Work

Beyond performing identities, media accounting becomes a way to *create representations* of identity work. My posts on Facebook to and about others are part of *my* self-representation. Media scholar Nancy Thumim makes the very helpful distinction between representation and presentation or performance:

> Performances of self, presentations of self and self-representations coexist and, of course, are all subject to the process of mediation. However, the precise notion of the representation raises questions about the mediation of a textual object. In this view, when a self-representation is produced it becomes a text that has the potential for subsequent engagement.[5]

Representation differs from performance and presentation because it involves the production of a textual object, such as a Facebook post. Thumim argues that the textual object of self-representation matters because it can be reengaged with. When people perform their identities through media accounting, they create representations. The media traces they write or create become texts which can be read by themselves and others. Thumim argues that the mediation of those traces matter because unlike performances which may be fleeting, representation is defined by mediation. Thumim's work on self-representations and digital culture therefore involves a mediation framework in which she examines public institutional, cultural, and textual forms of self-representation. This provides a helpful frame for situating media accounting and identities because it helps us recognize the ways that commercial, health, and religious institutions have long encouraged people to document their lives and the world around them.

Representations of Not Only Me but Us

From a dramaturgical or ritual perspective, representations of the self are never just about the self. Even the "selfie," which on the surface seems unabashedly individualistic, can be defined socially. Theresa Senft and Nancy Baym define a selfie as inherently relational as both an object and a practice:

> A selfie is a photographic *object* that initiates the transmission of human feeling in the form of a relationship (between the photographer and the photographed, between image and filtering software, between viewer and viewed, between individuals circulating the images, between users and social software architects). A selfie is also a practice—a *gesture* that can send (and is often intended to send) different messages to different individuals, communities, and audiences.[6]

As an object and a practice, selfies share meaning between people and socio-technical systems. While selfies can certainly be considered an act of personal reflection,[7] they must also be understood as a social act. From a dramaturgical perspective, the audience shapes the particulars of any identity performance. Therefore, the presumed audience helps to shape the practice and meaning of a selfie. But even more simply, the fact that the sentence "Let's take a selfie" makes grammatical sense suggests a dyad or small group engaging in a practice of photographic self-representation. The "self" in this case is the group—it is us. Group selfies and the intended audience reveal the sociality of the object and gesture.

Selfies have resonated in our contemporary culture. Indeed, "selfie" was the *Oxford English Dictionary* Word of the Year in 2013. Selfie culture seems to represent a fundamentally narcissistic culture, one obsessed with creating attractive images of ourselves. But what these arguments have missed is the fundamental social nature of selfies.[8] Indeed, group selfies help explain part of the allure of selfie sticks, those extendable poles that hold smartphones and allow users to take a photo. Selfie sticks expand the photographic field to more easily include other people in a selfie. Many public venues, however, including Disney theme parks and Smithsonian museums, do not allow selfie sticks, citing safety concerns that are often welcomed in the public discourse, which vilify selfie sticks as selfish intrusions into physical space.[9] Selfies and selfie sticks are often considered material representations of a narcissistic society.[10] Besides the social nature of selfies, these debates have also missed the difference between giving a stranger your camera versus your smartphone to take a picture. Compared to cameras, smartphones are incredibly intimate devices. We wake up and go to sleep with them. We often carry them with us wherever we go. They can hold financial information, business emails, or even intimate pictures. Therefore, it is much harder and potentially less safe to pass it to a stranger for a picture than it is to pass a camera. Selfie sticks solved the problem of getting more people into a shot without having to give up your personal phone to a stranger, if only for a moment. Selfie sticks allow us to more easily perform our sociality and create representations of our social performances.

Both selfies and selfie sticks belie the social character of this contemporary photographic practice. Long before we had selfies and front-facing cameras on our phones, media scholars argued that the identities of the photographers or scrapbook compilers are reflected in the media they create even if they are not the subject of the picture or scrapbook.[11] A comparison of snapshots, scrapbooks, and even location-sharing posts reveal the small ways that people perform their identities through media accounting of themselves and others.

The Rise of Snapshot Culture

Before Google and Facebook were the primary technology brands that people brought into their home, there was Kodak. George Eastman's Kodak No. 1 handheld camera, invented in 1888, fundamentally transformed professional studio photography into snapshot photography.[12] The development of Eastman's handheld camera and roll film fundamentally obscured the technical aspects of the photography from the amateur photographer. Jenkins argues that this decoupling of the photo-taking and the photo production process was the underlying force bringing about amateur photography at the end of the nineteenth century.[13]

Amateur photography can be understood primarily as a media practice rather than an artistic endeavor. While photography as a mode of artistic expression has dominated visual studies literature,[14] amateur photography has historically less often been the focus of intellectual inquiry despite its prominence in society. Indeed, throughout much of the twentieth century, photography by default was professional and *amateur* photography had to be signaled.

> As we look back on the history of photography, reciting and rediscovering the names of visionary men and women who have shaped the medium, it is easy to forget the vast majority of photographers remain nameless. Amateur photography, although it probably accounts for the greatest volume of work, is considered unimportant. Most of it, after all, is made and seen only within small circles of family and friends, and it's all just snapshots ... The value of amateur photography rests in the fact that it is photography by ordinary people who have made the pictures first and foremost to be seen by family and friends.[15]

The importance of amateur photography, like other forms of media accounting, reveals the communicative and symbolic functions of media in our everyday lives among ordinary people.

The early imaginaries of amateur and snapshot photography shape everyday media accounting practices even today. The slogan for the original Kodak camera was "You press the button—we do the rest," a phrase that encapsulates much of the snapshot phenomenon.[16] Much like the selfie today involves not just a specific genre of photograph but a series of morally evaluated practices of media accounting, so too did the snapshot at the turn of the twentieth century. The entry of a cheap and technologically accessible camera for the masses was met with derision by the professional photographers who sought to confirm their professional status as photographers. McCrum argues that photography is about creative expression, a technological as well as aesthetic expertise, while the snapshot is just a record.[17] Like other media accounts, the process of the snapshot is not necessarily a technological nor a creative one, though it may be. The process central to what makes a snapshot a snapshot is fundamentally a social process: "A snap-shot is consciously taken for physically showing to other people and talking about its contents."[18] Snapshot photography is meant to be shared and always has been.

Pierre Bourdieu's work on photography was one of the earliest explorations into the sociology of photography, particularly highlighting the social practices surrounding amateur and professional photography and the role of class in shaping what gets photographed.[19] Later, the works of Hirsch, Chalfen, and Zuromskis would also reveal the functions of photography within families.[20] Creating representations of the family or of specific family members through snapshot photography is a practice through which kin identities are performed. Photo albums and framed family photos both collect and display representations of the family. In early Kodak advertisements, the family and the home were also often photographic subjects conveying imaginaries about what snapshots are for.[21] Both the practice and resulting products of snapshot culture thus reify the family identity.

Media accounting is about not only the production of identity texts through photographs but also their consumption. Hirsch argues that her collection and assemblage of photographs of her grandmother are important for her own identity development and performance.[22] She understands who she is through her family's photographic media accountings. Identity work here occurs on multiple levels at different stages of media accounting. The identity traces of a grandmother are coupled with a granddaughter's own identity work in her collection and consumption of those traces. Here, identity work is as much about the account as it is about the accountability. As audiences, we are accountable to receive media traces, which in turn can become part of our own identity work.

The creation of media accounts ostensibly for others deeply involves collective identity work. The most prominent social collective is the family unit, whose identity is both enacted and displayed in media accounting. Most often women take on the role of representing the family through their collective media traces. Indeed, family historians are predominantly women who work to create and maintain representations of the extended family.[23] Media accounting that represents the family is part of kin work[24] or affective labor.[25] Family historians often seek to understand themselves through the tracing of their own ancestry.[26] Snapshots and their collections have long been an important mode for media accounting and play an important role in understanding and representing our own identities and the identities of others.

Aesthetics of Snapshots

Despite their diversity, certain visual aesthetics dominate early twentieth-century snapshots in middle-class white America. Snapshots and amateur photography were dominated by images of domestic and family affairs. Becker describes how the late nineteenth-century image composition copied the aesthetics of professional portraiture:

> The subject's face is centered horizontally in the frame, about one third the way down from the top. The camera lens is set at eye-level, giving a "natural" perspective. Illumination from above and from one side gives a sense of volume. Due to

the slowness of early film emulsions, photographers often placed the subject in an armchair, with a headrest giving additional vertical support. Even when faster film made this technique outdated, the aesthetic was not, and the composition of the portrait changed very little.[27]

Even when photos became less formal and varied their settings in the twentieth century, the symmetry, frontality, and composition of the primary subject did not change much within snapshot photography.[28]

Snapshots also seem to capture quotidian, sometimes mundane aspects of domestic life, and yet they both create and embody personal meaning and affective response. "Scholars seem to agree that one's personal attachment to a snapshot manages to supersede the image's banality with some form of deeply subjective meaning, culminating in an intimate bond to a particular image, and an intense pleasure and satisfaction in the viewing of that image."[29] For example, I posted a photo to Instagram on Mother's Day not of me and my mother, as is often the case on social media posts on Mother's Day, but of my son and my husband (figure 3.1). They are sitting

Figure 3.1
Photo of author's son and husband, May 8, 2016.
Source: Author's personal collection.

at my mother's kitchen table (the table is not visible in the photo but the kitchen cabinets are). My eighteen-month-old son wears a green sweater vest with subtle drool marks down the front, while my husband wears a blue button-down shirt and a pink tie, which has been loosened at the collar. My husband has one arm around my son and the other outstretched in front of them as if he were taking the photo himself. However, I know his arm is only outstretched in an effort to get my son to look toward the cameraphone, which I was holding.

In many ways, the image of my husband and son is what Zuromskis calls an "ideal" snapshot image whose goal is to depict the subjects positively.[30] Both subjects look good and are dressed up and smiling. While not stiffly posed (an impossibility for my son, in any case), they are both looking up toward the camera and seem happy, as if I just caught this moment of familial bliss by chance. Rather, it was one of six photos I took trying to get both my son and my husband to look at the camera and smile at the same time. The other five remain on my iPhone and only the best one got an Instagram filter and was uploaded to the web. But I didn't delete the other five; instead, I uploaded them to my computer when my phone's memory got low.

Despite me not being a subject in the photo itself, this image is a form of self-representation for me, particularly as *I* posted it on Instagram rather than my husband. My own identity as a partner and mother are reflected in this image. To further explore the sociality of my self-representation, it can be a helpful to explore whom this snapshot was taken for. For me? My husband? My son? My mother-in-law who loves photos of her son and grandson? My followers on Instagram who liked it? All of my imagined audiences helped to shape the identity performances at work in this image and must be taken into account when understanding its broader meaning.[31] While the personal nature of snapshot photography has been well-documented, the social and communicative function cannot be overlooked.

Social media platforms make the communicative nature of such self-representations increasingly transparent and networked.[32] The various cues and metadata associated with our posts are central to meaning-making and circulation processes. When we post snapshots of certain events on Facebook or Instagram, our social relations with those who will see the photos anchor the meaning of those photos. In tagging photos or writing messages to people through social networking platforms, we reify and display our relational status both to them and to the broader network.

Snapshot aesthetics allow people engaging in media accounting to perform various aspects of their identities. Most prominent is the focus on the domestic, but other themes also reveal themselves. Leisure and play were prominent themes in early amateur photography.[33] Travel and vacations were featured as particularly capturable by the Kodak camera. While this conveys a sense of leisure and class,[34] the theme of mobility is also an undercurrent to these advertisements. For example, a 1910 ad with the tagline "Kodak, as you go" depicts a servant bringing a Kodak camera out to a woman seated in the back of a convertible,[35] suggesting that travelers should be sure to bring their cameras along on trips. These ads depict leisure and mobility as an outside activity. This was partially because the camera could only take photos when the lighting was good so it necessitated outdoor light. But this also was in contrast with the professional studio photography. The playful and leisurely imagery in the advertisement conveys the kind of mobility that the handheld or pocket camera afforded. Much like the pocket diary afforded mobility, the handheld Kodak cameras at the end of the nineteenth century also used the mobility of the technology as a key selling attribute and improvement over its predecessor. Such mobility was represented in the aesthetics of early snapshots in complex ways. While most early snapshots needed outdoor light to create better images, the subjects of the photo still needed to remain still. So, there's an important contradiction in snapshot photography, which is the simultaneous need to remain still for a moment while also communicating some kind of act or action.

Over time, the key function of snapshot photography shifted slightly while maintaining a focus on the domestic. Nancy Martha West notes that, prior to 1900, Kodak advertisements primarily focused on playfulness, leisure, travel, and not having to worry about the technical development of film. After World War I, there was a major change in thought about the purpose of the camera and photography, shifting from leisure to "confirmations of family unity."[36] The identity work became less playful and more obligatory, especially for the female consumer.

Media Accounting and Consumer Culture

The growing consumer culture of the early twentieth century was reflected in snapshots and scrapbooks as media accounting modes of identity representation. The growth of the advertising industry at the turn of the twentieth century helped new consumers learn what kinds of identities were important to perform through their media accountings. Garvey notes that as early magazines shifted to an advertising-based model in the 1890s, the stories as well as the ads in the magazines construct the identities of consumers as female and tie the good consumer with the good middle-class wife.[37] Indeed, at the turn of the twentieth century, an increasing role for American women was that of the consumer.

At the turn of the twentieth century, in the midst of this growing mass media environment, snapshots as well as scrapbooks allowed people to create traces of their own lived experiences and in doing so it allowed them to perform various aspects of their identities as consumers. As West puts it, "to make the real consumable is to affirm it."[38] Creating and collecting snapshots or any media trace of our lives allows us to show off as well as consume these identity performances at a later point in time. "Scrapbooks represent individual and group identity in cultures increasingly dependent on reading, visual literacy, and consumption of mass-produced goods."[39]

While highly personal media accounts, snapshot practices are also tied to consumer culture. From photographing ourselves with cars and houses or in our new dresses or suits to snapshots on the beach on vacation, snapshots depict our consumer culture at home and at play. As Zuromskis writes, "As snapshots become ever more ingrained in the American cultural imagination, it is increasingly difficult to separate the way snapshots are taken, and indeed the very impulse to photograph, from the guiding interests of commerce and the culture industry."[40] Scrapbooks, perhaps even more than snapshots, clearly reflect a consumer society.

When people create and collect media traces of themselves and the world around them, they enact both the performing and the performances of identities. In their 2006 edited book, Tucker, Ott, and Buckler argue that scrapbooks are about both the individual identity as well as larger cultural and group identities. They can reveal aspirational identity as well as certain aspects of our lives, including our values and what is important to us. Media accounting is both the captured and collected identity traces that affirm particular identities, but the consumption of media accounting by the self and others reaffirms such identity work.

Representing the Self through Others' Media

The simplicity or the naïveté of the snapshot or scrapbook is also something that is central to media accounting. Often overlooked as a real or legitimate form of media use or production, the simplicity of media accounting, much like the snapshot or scrapbook, belies its larger role in identity work and broader cultural production.

Scrapbooks as Identity Work

Amateur photography was not the only popular form of media accounting at the turn of the twentieth century. Scrapbooks were also common means of people cataloging the world around them. Ellen Gruber Garvey defines scrapbooks as "the embodied practices or gestures of cutting, arranging, and posting materials, and displaying the resulting books to others."[41] Often using newspaper clippings and other free or cheap printed materials like invoices, letters, ticket stubs, and advertising cards, scrapbook compilers would both keep information and news of events of importance to them.

Scrapbooks varied in their personal and professional nature.[42] Some scrapbooks were highly personal, places to grieve the loss of a loved one or to imagine a better life for oneself. Other scrapbooks were much more strategic and political in nature, such as those compiled by politicians, suffragists, and abolitionists. Still others are highly professional, documenting the events and news coverage of an organization or individual professional such a doctor, lawyer and even actor. As Ellen Walkley (2001) notes, however, the clear distinction between public and personal scrapbooks are blurred. Social and civic clubs also kept scrapbooks of their activities and development, as did families and churches. Individuals would keep scrapbooks of their professional, civic, and personal interests. The variations in content are so great among scrapbooks that it is most accurate to unify them in practice rather than content, defining scrapbooks as the compiling of materials into a new form.

One of the key defining characteristics of scrapbooks is that scraps are arranged and displayed in a book form. It was common in the late nineteenth century to reuse books and paste clippings over other books when commercial scrapbooks or albums were not available.[43] But displays are more than the compilation of materials; they imply an audience. Sometimes the audience is merely the future self. Sometimes this audience is future generations within the family. The beautiful and ornate nature of scrapbook covers also suggested that these books were meant to be shown off to others.

Garvey suggests that part of the reason why scrapbooks became an important cultural practice was due to the proliferation of inexpensive material forms. The rise of color printing and the availability of scraps of color paper fueled scrapbooking practices in Europe and North America. The dominance of commercial products or materials within scrapbooks has led them to be dismissed as cultural forms of expression and production; however, these books were immensely meaningful not only for the compilers of them but for those historians, anthropologists, and sociologists who seek to understand individual as well as collective practices and meaning making.[44]

The meaning making in scrapbooks through the compiling of materials has been described by Garvey as scissorizing, a mode of composite authorship. She argues scrapbook makers "told their own stories in books they wrote with scissors."[45] The recontextualization and recirculation of clippings and other printed materials via scrapbooks becomes an authorial act. By selecting what to include and what to leave out, scrapbook compilers create a culture remade for one's own use. Thus, it is both the collection and curation of material forms that give rise to the value of scrapbooks for identity work. Like forms of commonplace writings,[46] ordinary and everyday forms of media production like scissorizing reflect individual as well as collective social identities and values.

The identity work within scrapbooks is important because it reveals the way that people can present various aspects of themselves through materials created by others. Unlike the diary and amateur photography where people are creating media traces, the scrapbook represents a mode of identity work wherein the author is using media created by others to engage in their own media accounting. As Garvey argues, this mode of identity work represents a cultural context in which people feel a kind of information overload.[47] There is so much content available to them that they do not need to create their own. They can create meaning and engage in performances of the self through the compilation of content created by others.

The identity work and meaning making occurs in both the choice as well as the arrangement of media "scraps."

Scrapbooks occasionally included the written thoughts or drawings of the compiler. When people would clip articles from newspapers, sometimes they would also include their thoughts or opinions on the articles as well. For example, Ellen Walkley (2001) found that George Himes included the newspaper clipping entitled "Ford May Save the Day: Detroit to Give His Street Car Real Tryout," 1920 with a handwritten note: "Something must be done to solve the city transportation problem. The above Ford project may lead to the solution of a solely vexatious questions. The Detroit experiment [illegible] closely watching."[48] Another example was the clipping "Rich Have Nothing to Do but Work Due to Dry Laws: Printer Worth $200,000 Back at Old Job to Pass Time," February 7, 1920, which included the following handwritten note: "What a life? Where there is so much that needs doing. A man like the one above described is a barnacle."[49] These examples reveal that scrapbook compilers were not only collectors and curators but commentators as well.

The combinations of clippings and commentaries are reminiscent of the kind of postings we regularly see on Facebook and Twitter today. In a content analysis of Twitter messages, we found that it was very common to share news or information available elsewhere on the web and to comment on it.[50] This is exactly like we see George Himes doing in his scrapbook in 1920.

Identity Work on Pinterest

Just as the turn of the twentieth century saw a proliferation of inexpensive material forms, I would suggest that the turn of the twenty-first century saw a proliferation of inexpensive but high-quality digital images online. Pinterest is a social networking site which facilitates the collecting and sharing of images—what they call "pins." Unlike other image-sharing sites, most people do not upload their own images to Pinterest but instead tag, collect, and curate images from within the site or those found elsewhere on the web. Like scrapbooks, Pinterest is a site for people to collect images and other bits of media created by others and assemble them in a way to make personal meaning. In doing so, they reflect various aspects of their own identities in several ways.

In our study on Pinterest, we found that people's creation, curation, and consumption were important for the identity work.[51] The people to whom we spoke were thoughtful about the ways their user profile reflected their identity, which included not only their username and profile location and a biographical blurb.

In addition to the general profile information, identity performance was also central to how people in our study pinned. The act of pinning on Pinterest is fundamentally about collecting pins. What people choose to collect pins *about* and what specific images they pin are ways of representing various aspects of their identities. Foodies pin recipes and images of food. Crafters pin how-tos and images of successful crafts. In our study of designers, they were very careful to only pin images related to design, mostly interior design. The pins were each beautiful, representing a particular aesthetic or point of view of the designer himself or herself. Both the particular pins as well as the theme of the pins chosen represent the identity performance or performances.

It is not just the pins and their topics that represent the pinner's identity, however, but how the pins are arranged on the boards also conveyed important aspects of our participants' identities. The ordering of the pins is like the telling of a story. Arranging different pins next to each other conveys different information. The juxtaposition and aggregation of images on a board allows them and their followers to see different things. The designers in our study were well aware of this and strategically arranged pins on the boards to convey their design aesthetic. The arrangement of the boards was also central to their identity work as designers. Much like the arrangement of scraps in a book tell different stories, so too do the arrangement of pins on a Pinterest board.

Identity work can also be performed through whom people choose to follow on the Pinterest. The designers in our study were again quite conscious to follow people who reinforced their identity as designers, so primarily others in the design community. And like other social network sites, who follows you is also an indication of your identity.[52] The designers in our study prided themselves on having a lot of other designers follow them as well as having lots of followers in general as an overall testament to their expertise as designers and thought-leaders in their field.

Baby Books as Identity Representations

Baby books are a form of media accounting that showcase the point that media accounting is not only about the author or writer but also about and for others. Mothers are often charged with creating media traces for their children because they cannot yet account for themselves. Baby books did not emerge as a genre until the late nineteenth and early twentieth century. Janet Golden and Lynn Weiner argue that this is likely due to the dramatic decrease in infant mortality. They define baby books as "an item with one or more printed pages for recording information about newborns."[53] While some were quite fancy hardbound books, others were merely pamphlets with advertisements and only a few sections to enter information about the baby. Most often women, mothers more specifically, keep these books on behalf of their children. Golden and Weiner suggest that baby books were primarily for the recording of gifts and who they were from, chronicling of developmental milestones, including both height and weight as well as baby first steps, etc. Baby books also document religious events, such as baptisms and confirmation, as well as holiday rituals such as baby's first Christmas or Easter. Last, baby books also preserved photos, locks of hair, birth announcements, and palm or footprint tracings.

Baby books in the early half of the twentieth century also reveal the rise of commercial culture. Many commercial baby books were free to new mothers and included pages of advertising and medical advice. Not only did they advertise products that were for children such as infant formula and clothing, but insurance companies were common advertising sponsors for these books as well. The books also convey a focus on health marketing. Golden and Weiner write:

> The cultural practice of recording an infant's life in a baby book merged seamlessly with the public health command to record growth, development and medicare care. *Infant Care,* the federal government's publication for mothers that began distribution in 1914 (and was revised and reprinted numerous times, with millions of copies in circulation), contained a page for record keeping.[54]

Thus, baby books served the important role of keeping track of the developmental milestones, quarantine cards, and immunizations of children but also convey the merging of infant care and mass consumption.

The earliest baby book in the UCLA Louise M. Darling Biomedical Library archive is from 1882, entitled *The Mother's Record of the Physical, Mental, and Moral Growth of Her Child for the First Fifteen Years*. There are three important points that these books convey. First, the advertisers and manufacturers of these books presume or hope that mothers will be chronicling the lives of their children for a long time. Fifteen years is a long time to keep any book, let alone to maintain the act of accounting throughout that time. Second, books would come with prompts, pages to be filled in. Important milestones are already articulated and just need to be filled in with details, such as the day and characteristics of the events. For example, baby's first tooth appears with a section for notes about it. Another important point about baby books is that they tend to be incomplete. They were most often only written for first-born children and then only partially filled out. Pages were left empty. In many ways, these books set up mothers to fail.

Unlike a race which signals a clear start and finish, the most "important" childhood developments are processual—not discrete moments or events.[55] From first teeth to the first steps, most of these developments occur and are experienced as a process of maturation. Is the date we are to record a baby's first tooth the first time we can feel it under the skin (the baby is definitely experiencing the tooth at this point) or is it the point at which it breaks through the gum? But then it's really only a bit of tooth, not a full tooth. Similarly with steps, learning to walk is such a developmental process. One could easily put any date within a month to "accurately" capture a baby's first steps. It's not as if one day the baby is crawling and the next they walk around the room. It occurs in minor little breakthroughs; ones so minor we sometimes don't even acknowledge or see them. Yet the identity of the mother is necessarily reflected in the baby book she creates for her (first) child. Parents, and mothers in particular, are often charged with creating media traces for their children because they cannot yet account for themselves.

Historical changes in how mothers describe their child's activities and accomplishments also reveal the increased awareness of how media accounting reflected not just the child but the parent as well. Golden and Weiner found that older books were more likely to document the mishaps and missteps of young children, for example, babies falling down stairs or off furniture. However, it's unlikely that young children stopped falling down; instead more likely mothers merely stopped documenting it.

Golden and Weiner suggest that when infant mortality was more common, the vitality of the child was not seen as a reflection of the parents, but, as infant mortality decreased, the vitality of children was increasingly associated with parental identity.[56] The media accounting on behalf of the babies reflected this growing awareness linking parental and child identities. Historically, the practice of baby books emerged as part of a public health initiative, but today social media prompt us to create traces of ourselves and others every time we open them.

Today there are online services like "My Own Little Story: The online baby book that won't let you forget" to help parents document the lives of their children. They are clearly targeted at busy and sleep-deprived parents who mean to document their child's milestones. The service sends email reminders to parents about particular developmental milestones worthy of note, such as first smile, favorite foods, sitting, favorite games, and first words. The service, like many other baby books, encourages the parents to write about the baby shower, other forms of preparation for the baby, as well as how the parents are feeling about the upcoming birth of the child—further blurring the identities of the media account. My Own Little Story is not just an online service, but parents can pay to have it printed into a book. The website is full of photographic examples of physical baby books. The printing of the baby book is an important part of the media account because typically books are given to the child when they reach adulthood. While digital goods are giftable, the inability of two people to have the same material book makes its gifting seem more valuable.[57]

Snapfish is another service that allows parents to make photo books; it takes their digital photos and creates a themed book around five designs: family, travel, wedding, baby, and everyday. Representing middle-class media accounting genres, these five designs include templates, color schemes, and example pages to help mothers create gifts and keepsakes for the family. The digital and the print-on-demand nature of these books enables easy gifting of these books, in ways that were much harder with analogue baby books and photo albums. Indeed, Snapfish frequently has deals such as "Buy one, get one free." Like most digital media,[58] the work is in creating the first book, and subsequent books are easy to print and give away. The ability to print two of the same media accounting book suggests that the mother could keep a baby book for herself, recognizing how the book represents her as much as her child.

The Work of Representing Identities

Despite the fact that snapshots, scrapbooks, and even baby books are seemingly easy and fun ways for people to use media to document their lives and share them with others, I want to argue that the representation of identity work is a form of labor. Media accounting for the self or family could be identified as domestic material labor in that people are often producing representations on behalf of themselves and the family.[59] It is only in the contemporary moment that this practice has shifted to the digital. Long before Facebook and Instagram, mothers have been carefully creating representations of their children and families. Indeed, scrapbooks were long overlooked authorial acts that showcased the identity, politics, and desires of the scrapbook compiler through media created by others.[60] Because the authorial act only reveals itself in the aggregated curatorial form, scrapbooks were often not considered forms of media production.

The supposed easiness of performing identities through media accounting hides the vast amount of resources than can go into the practice. Indeed, people can spend hours a day "working" to create these traces in their aggregate form. Whether it be the scrapbook or the photo album or the baby book, compilers are always encouraged to do more. The work of representing our identities in media is always incomplete. Much like housework, it is never done. There is always more information than can be represented both because the content is ubiquitous and because social media platforms are always asking for it.

Liesbet van Zoonen and Georgina Turner argue that much of the identity work that goes into media production has shifted from identity expression and moved toward identity management.[61] The key distinction between the two is agency. Identity management is less empowering and necessary for interactions with commercial and state authorities, whereas identity expression is a long-standing practice in self-narratives and storytelling. But the division between management and expression is not clear. Jefferson Pooley argues this is a bind between representing ourselves for both self-fulfillment and self-improvement, something he calls "calculated authenticity."[62] Thus, I post only the best of the six photos of my son and husband.

Media accounting has long relied on commercial products to represent identities. The key shift is not in the commercialization of identity representations, but in the ownership of representations. Even within digital identity representations, we see increasing material production. Services like Snapfish and Shutterfly will print and create images and books of our media traces. We see digital identity representation being turned into coffee-table books, mugs, T-shirts, hats, etc. to be exchanged or gifted among close relations. The creation and exchange of identity presentations in media accounting can be considered part of "women's work,"[63] as it both fulfills relational expectations among friends and within the family and represents the relations as well.

Collective Identities and Context Collapse

One of the primary tensions identified in social media research is "context collapse" or the inability to distinguish audiences for one's identity performances.[64] It is argued that social media blur contexts for communication, disabling people to tailor their identity performances to particular audiences. When I posted the photo of my husband and son, I was not thinking of all my professional colleagues on Facebook, but those who know my family. Van Zoonen and Turner argue that many digital media platforms insist on a singular identity, despite the fact that most cultural and social theory understands identity to be multiple, dynamic, and contextual.[65] For example, Facebook only allowed people to create one profile that was supposed to represent a particular offline identity, by insisting people register with .edu email accounts. There are numerous stories of people losing their jobs after posting off-color remarks or images intended for personal friend networks, which became publicized within a public or work context.

Over time, user strategies have mitigated potential social faux pas that may emerge from having multiple audiences, most commonly the tensions between postings for work colleagues and personal friends.[66] For example, people on Twitter or Instagram set up different accounts on one platform to manage these audiences. Sometime people use different platforms for different audiences. Sometimes they will just keep the content of their posts benign enough so as not to offend friend, follower, colleague, or foe.[67]

Most social media platforms have responded to these grassroots strategies by discouraging users from registering with different email accounts. By enabling grouping of recipients as well as multiple usernames or profiles to be associated with one email account, these companies at once acknowledge the multiplicity of identity roles that many of us play and ensure proper user identification, which is necessary for increased personalization and targeted advertising.[68] As long as social media platforms can keep us within their walled gardens, the identity work that we engage in can be packaged and sold back to us.

Summary

This chapter explores the ways that people use media accounting to create representations of identity performances. I want to argue that these representations are always about the self *and* others. Clearly, delineation between the two is nearly impossible. Whether it be the presumed audience or people who are subjects of the presentations, the roles we play are typically relationally defined and the representations we make reflect this.

Another important argument in this chapter is that people engage in identity work through the media of others. The "making" or producing of media does not have to be the primary mode of identity work. Instead, people can create representations of their identities through the scraps or snapshots or even tweets of others. Therefore, it is also in the collection, curation, and consumption processes that our qualified selves are represented.

People do not have to feature into the content of their media traces to reflect the qualified self. Baby book and family photo albums are prime examples of the ways that traces of others figure into our own sense of self. Posting photos of others on our social media accounts can reflect our relational selves as much as photos in which we are the subjects. The qualified self emerges not only in representations of ourselves but also in representations of our relations in our media traces.

The identities that we create representations of are not just actual, but aspirational as well. We want to put our best face forward.[69] Sometimes the representations that we create or curate reflect our hopes and dreams more so than our actual life, but they nevertheless represent us. Our aspirations sometimes say more about who we really are than our lived experiences do. The qualified self is shaped by representations of who we are as well as who we want to be.

Last, this chapter shifted the focus from identity representations to identity work as a form of labor. The time and effort that goes into creating representations of identity can be significant, despite discourse surrounding these modes of representation. The work of producing representations of identity is not necessarily immaterial labor as is characterized by much white-collar or digital production. Often material artifacts are produced as part of the representation process such as scrapbooks or photo albums, but they are also gifted, exchanged, and passed down within the family or even beyond as social media platforms enable. The immaterial labor of kin relations within media accounting can turn into the material labor of creating and maintaining familial traces over time. The material representations of media accounting can become valuable keepsakes representing our qualified selves.

4 Remembrancing

Media accounting enables the remembrancing of experiences, people, places, and events. I purposefully use the term "remembrance" rather than remember or memory because it focuses our attention on the active, social, and reflexive processes of memory. A remembrance is a reminder, but the practice of remembrancing through media accounting is the creation or use of media traces as part of our memory work regarding ourselves, the people in our lives, and the world around us. In this way, remembrancing is not just static cognition, but something one actively does and engages in. Fundamentally, remembrancing is about creating and engaging with media traces to help us remember.

We regularly post on social media about events or experiences that we anticipate we will want to look back at and remember. Creating media traces is a way of holding on to experiences. Posting photos from holidays or vacations allows us to revisit and share these events with others. Sometimes a holiday photo is the only static memory trace of an otherwise chaotic experience. Travel and vacations are typical experiences that remembrancing enables the documentation and reliving of. We often purposefully create traces in anticipation of future use.

Memory Work and Mediated Memories

Remembrancing is a form of what the memory scholar Annette Kuhn calls "memory work."[1] This is an active process of remembering that questions the implicit authenticity of objects that hold memories, like family photographs or souvenirs. Kuhn argues that memory work is a conscious process of staging memory. While memory might seem unconscious, memory work highlights the ways that people purposefully and strategically create media traces to help them remember events and experiences in their lives within particular narratives of the self, the social context (e.g., the family or romantic partner), and the broader cultural environment. Therefore, remembrancing as a form of media work intertwines the personal, the private, and the public.

Media play a vital role in memory work. The media scholar José van Dijck uses the term "mediated memories" to convey both "the activities and objects we produce and appropriate by means of media technologies, for creating and re-creating a sense of past, present, and future of ourselves in relation to others."[2] Here media and memory are mutually shaped and constituted. Media technologies, like photo albums, scrapbooks, or Facebook, are not merely memory objects that hold or store our memories for us. Instead these media actively shape our memories, for example, reminding us how birthdays and weddings should be captured and saved.

Christine Lohmeier and Christian Pentzold extend van Dijck's understanding of mediated memories to argue that mediated memory work are "bundles of bodily and materially grounded practices to accomplish memories in and through media environments."[3] Mediated memory work is not just cognitive but somatic in nature. They argue it simultaneously helps people maintain a sense of individuality while connecting them to larger collectives and communities through both cognitive as well as emotional activities. Mediated memory work also has a materiality, whereby media technologies become the means through which people "work on and with memories."[4] Remembrancing is the mediated memory work of media accounting.

Remembrancing through Media Accounting

Remembrancing is creating a media trace about an experience that allows us to hold on to moments or experiences. When we write in a diary or a tweet about something we just saw or experienced, this helps us to remember what it was and what it was like. Indeed, the act of creating the trace, whether we ever look at it again, helps us to remember things we have experienced or heard.[5] But we don't create media traces of everything we hear or see. We strategically choose certain things and not others to create remembrances of. Our purposes for remembrancing may involve other media accounting practices such as performing identities. For example, I may post that I went to a Radiohead concert to align my identity with their musical genre but also because I don't want to forget the experience. It's something I want to hold on to.

The creation of a media trace, whether it be writing the text, taking the photo, or writing the text about the photo, invites reflexivity. This can be both explicit or implicit. When I take a photo of my son on his first day of full-time day care, I am implicitly calling attention to the fact that this is a moment of transition in his and my family's life. When I post the photo with the text that this is his first day of day care, I make explicit the context for posting this particular photo, so that others can implicitly understand the social significance of the image, that is, parent/infant separation. Otherwise, the photo is just a depiction of a baby in a car seat outside of a classroom with toys.

Media accounting is common during trips and vacations in part due to remembrancing practices. When events are new or unusual we often want to remember them; therefore, we are more likely to engage in remembrancing to help us hold on. Travel journals or blogs are very common ways that people engage in remembrancing. So even if one does not tend to keep a diary or to post often on Facebook, travel becomes a moment that might motivate someone to engage in media accounting. For example, there are even specific journals dedicated to chronicling travel. Diaries such as the one depicted in figure 4.1 from 1952 can be particularly formatted for travel remembrancing. This is not a blank journal, but one designed with several sections explicitly related to travel. The first section includes twelve pages of text about traveler tips such as how to pack baggage, how to get foreign money, European time changes, and what to wear.

This section of the travel diary also includes specific tips regarding steamship travel, like definitions of deck sports, nautical terms, ship's times, fog signals, and what to do in case of seasickness. The majority of the pages of the diary are lined pages entitled "Events and Places visited." On the top three lines of each page are printed "Date," "Place," and "Weather." Sometimes the author would fill these in, such as: "Date: May 23 Place: Rome Weather: Sunny A.M. Rainy P.M. I have seen so much and heard so much today it is almost impossible to write about it…" Of course, he tries to capture his experiences in the next six pages as he describes going to museums and what he saw. In the back of the travel diary is a section called "Addresses," with an illustration of a US mail box and a bird carrying an

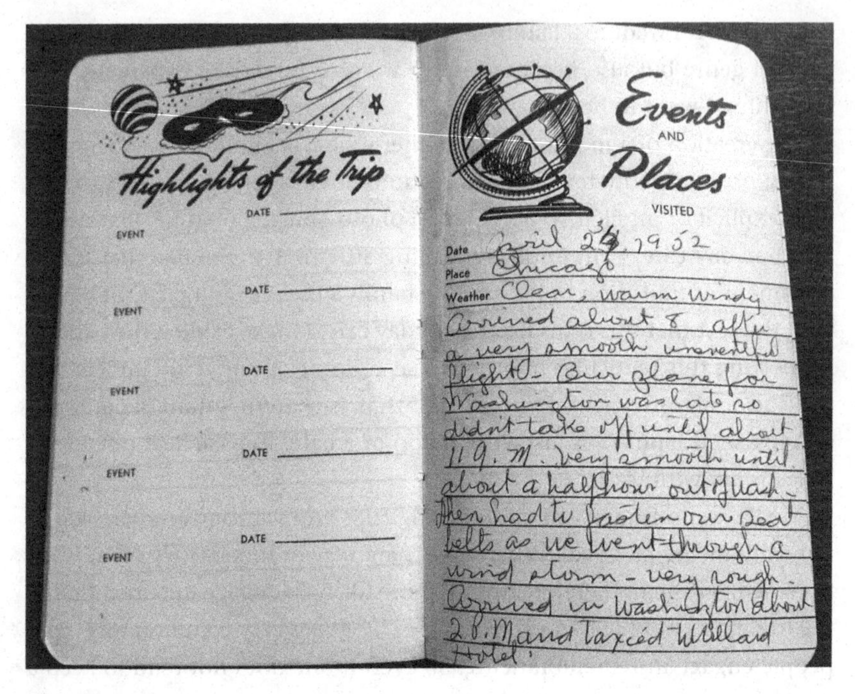

Figure 4.1
Travel journal from 1952.
Source: Author's personal collection.

envelope. Indeed, six of the twelve pages are full of American addresses in alphabetical order. Travel diaries or journals help keep track of one's experiences as well as the ability to share these experiences with friends and family at home. This not only reinforces the relationship but also helps those back home to know that one is still safe. Esther Milne calls these technologies of presence, whereby the communication "enable[s] complex exchanges, experiences as the intimate presence of a correspondent whose corporeal body remains invisible."[6] Remembrancing is not just through the creation of media traces, but the sharing of them as well.

When I began this book, a new social media platform had emerged, Snapchat. The service allows people to take "snaps" of themselves; decorate, annotate, or manipulate the images; and then share them with others. Many people take snaps of themselves making funny faces or doing funny things, or just take pictures of things they encounter in their everyday life. This is really the key to media accounting—to documenting oneself or the world around us and sharing it with others. However, snaps disappear after you open them. Therefore, while the other practices of media accounting were certainly relevant to Snapchat, remembrancing didn't seem to be among the practices achievable through the service. Indeed, while technologically many social media platforms are made for storing and collecting traces (and data), such accumulation was not part of Snapchat. In trying to explain the service in June 2015, Snapchat CEO Evan Spiegel said,

> Snapchat really has to do with the way that photographs have changed. So historically photographs have always been used to save really important memories, major life moments, but today with the advent of the mobile phone and kind of this idea of a [air quotes] connected camera [air quotes], pictures are being used for talking. So when you see your children taking a zillion photographs of things that you would take a picture of, it's cuz they're using photographs to talk. So you can think about then photographs being really about saving up memories. Now, photographs are really used for talking. And that's why people are taking and sending so many pictures on Snapchat everyday.[7]

According to Spiegel, photographs and other social media platforms are for remembrancing, but Snapchat is for talking. However, as people engaged in the creation of media through their snaps, they wanted to keep them. They would take screenshots of their own and others' snaps as a means of holding on to them. In July 2016, Snapchat revealed its Memories feature, which allowed users to keep some of their own snaps. Sometimes people create or receive snaps they wanted to keep and look back on. The default is not to store a snap, therefore there is an anticipatory nature to adding a snap to one's memories. One must save a snap before sending it. Taking a snap at a party or graduation and saving it suggests wanting to keep or hold on to these events. Other times as people play with Snapchat features and filters they engage in mediated memory work, wherein the act of creating the media trace is what creates the memorable moment. Using a funny image filter that makes your friend look like a dancing hot dog is a way of displaying and creating a remembrance of your social bond. The act of annotating a snap further can facilitate the remembrancing, by contextualizing or creating a metanarrative for the image.

Snapchat has become so commonly used for communication that people also use it to document various life events. Sometimes, people use Snapchat to communicate about events or experiences that they wanted to share *and* remember. Therefore, remembrancing became part of what Snapchat and other media accounting platforms enable users to do.

Understanding the Rosalyn Racca Incident

On August 29, 2015, Rosalyn Racca shared photos with her friends and family of her son, Samuel Tate, on the one-year anniversary that she delivered him. Like many mothers, Rosalyn posted a series of photos on Facebook remembrancing the event. However, the photos were reported as inappropriate content and Facebook asked Rocca to remove them or change them to private, else her account would be suspended.

The reason the photos were flagged is because Rosalyn's son Tate had been born dead.[8] A routine check at twenty-one weeks of gestation revealed that the baby had died. She delivered the baby the next day. Photos were taken after the delivery of Rosalyn, her husband, their other children, and Tate. In one news report Rosalyn Rocca is quoted as saying, "The whole reason I was sharing this was to celebrate his life and he was my son. That's all I see when I look at his pictures."[9]

Photos of prenatal or infant death may seem startling or unusual. However, it is not only recommended by contemporary medical and psychological professionals, but it is also a long-standing media accounting practice. As Burns notes, "In modern Western culture, to have no photographs or visual representation at all is not to have existed. A child touches the lives of its parents no matter how long it is with them."[10] The media traces of people, often through photos but also through other media, is an essential way that we remember them.

When I was researching mid-to late nineteenth-century photographic practice, I came across postmortem photography, in particular infant postmortem photography. Infant mortality was still relatively high during that period, so often families would not have had a chance to photograph the newest member of the family before he or she died. Therefore, it was typical to photograph young children right after they died. Called memorial photography, these images were a common middle-class practice in the US and Europe in the mid- to late nineteenth century and into the early twentieth century.[11] In most cases, the child would be photographed as if he or she were sleeping, either held by a parent or lying on a blanket in a cradle or even a casket. Rather than "sleeping," some memorial photographs depicted the dead child with their eyes open, or the photographer would paint on eyes or add color to the cheeks during the development process. The babies were often in dressing gowns or wrapped in lovely blankets and sometimes photographed with toys. Most of these images are black and white.

By the mid-twentieth century, memorial photography was taboo; by that time, death primarily occurred not in the home but in the hospital.[12] Undertakers and hospital morgues began to professionalize the preparing of the body for burial or cremation. No longer the responsibility of family members, death was no longer part of everyday life as it had been in the nineteenth century. It was common practice in the mid-twentieth century to take stillborns or very ill infants away from their parents immediately after delivery so parents would not have to see their dead children, which was understood to be overly traumatic.[13]

In the early twenty-first century, however, we have begun to see a shift in practices around death. Today, the Western medical and psychological communities understand the importance of parents being able to hold and spend time with their infants after they have died. Indeed, it is now standard care to offer the baby to parents to hold, as well as to offer professional photography and footprinting in situations of perinatal death.[14]

Whether you call it bereavement or memorial photography, the photography of death, particularly of infants, is an important remembrancing practice. The most popular organization in the United States advocating for memorial photography is the Now I Lay Me Down To Sleep Foundation. The organization uses the term "remembrance photography" and is operated as a nonprofit where parents do not pay for professional photos. Parents of "babies who are stillborn or are at risk of dying as newborns" can get assistance from the foundation. The foundation pairs parents with professional photographers who will come to hospital and take professional photos of the parents and their baby free of charge.

The examples of remembrance photos on the Foundation's website are reminiscent of the nineteenth-century memorial photos. Always in black and white or sepia-toned, photos often depict the mother holding the wrapped infant with the father close by, both parents looking lovingly at the child. Sometimes the photo is of the baby alone, looking as if he or she is sleeping peacefully. One testimonial on the Foundation's website said, "We looked at [the photos] right away because we needed, we wanted, photos for the service. We wanted people to see that yes, we had a daughter, and yes, she was beautiful and perfect. And we wanted to share her with the world." Retouching photographs was a common practice in memorial photography in the nineteenth century and has only become more advanced. No longer a duty left to the photographer, the foundation hires professional digital retouch artists who can "repair skin flaws (torn skin, blistering)" and "even out skin tone (bruising, discoloration)." The resulting images are well-lit, beautifully composed professional portraiture.

In terms of the composition of memorial photos, there are similarities and differences between the nineteenth-century photos and the contemporary remembrance photos taken by the Now I Lay Me Down to Sleep Foundation. The images are often of the dead infant as well as the dead infant with the family. The Victorian dress of grieving parents was often suits and dresses for the memorial image and formally posed at the home of the family or in the photographer's studio. Today, the photos are nonetheless posed, but are often taken in the hospitals. This is because a hospital is both a place where children are born and a place where death happens, and this duality impacts the composition of the photos. For example, today sometimes parents are still wearing hospital gowns in the photos but may have a blanket over them to make the photo less clinical. Today the infants themselves are more likely to be depicted naked or just wrapped in a blanket. Like the nineteenth century, however, today parents or nurses will often dress infants in white baptismal or dressing gowns or include keepsakes or toys from older siblings or family members for the photographs. The composition of the resulting images conveys not just the existence of the infant in this world, but his or her familial presence and relations as well.

Another key difference between nineteenth-century memorial photography and contemporary images is the closeness of the image frame. The foundation's photos are typically very closely framed, sometimes with only the faces of the family or baby depicted, suggesting a sense of intimacy. Nineteenth-century photos of parents with their deceased children would be taken from five to ten feet away to properly focus the image. Even when the image is just of the baby, historical images typical depict the entire body of the child. If it does depict just the face of the child, it still looks as if it was taken from about five feet away. Today, the images look as if they are taken only inches away. The up-close nature of the images captures the tiny features of the babies. The contemporary remembrance photos sometimes juxtapose the parents' size with the baby's size, depicting the hands of the baby and parent together to really capture just how small the baby is. These contemporary images highlight the small, delicate bodies of these babies. Clear aesthetic and compositional differences between contemporary and nineteenth-century memorial photography can be understood as reflecting both the changing technological abilities in cameras, as well as social conventions regarding birth, death, and displays of intimacy.

While postmortem media traces of children may seem morbid, they can help grieving parents. Today prenatal and perinatal mortality photography is understood as important to the grieving process and recommended by grief counselors and obstetricians.[15] Footprints and photographs of infants are common ways that hospitals will create lasting traces of a dead child for parents to hold on to. Sometimes parents won't look at the media traces until years later, and sometimes the photos used at memorial services are an active part of their grieving process when they leave the hospital, helping to make real what can be a surreal experience. Another parent on the Foundation website said, "These helped me a lot because my baby exists. She's right there. She's in the pictures and friends and family that go over to my house, you know, this is my baby. She lived. She was here with me. I carried her and here she is." These media traces help parents to have material artifacts as proof that the child was part of their life.

Remembrance photos may become cherished by families who have lost someone too quickly. They are particularly important in prenatal and perinatal mortality because the families did not get a chance to create media traces of their child alive or healthy. Therefore, the postmortem photos are the only media traces these parents will have, outside of sonogram photos. In instances where the infant lived for days or months, postmortem photos may still be highly prized because they can depict babies who look like they are sleeping rather than babies who are sick or hooked up to hospital monitors and struggling for life. Indeed, sometimes one cannot tell the difference between a postmortem newborn photo and one of a baby that is healthy and alive.

Contemporary remembrance photos can also be understood as part of identity work. The identities involved in the posting and sharing of remembrance photos may be multiple. Primarily, these photos represent parental and familial identities. The photos represent a child in the family who died. But the images also represent the identities of the parents. The photo is part of their own identity work of what it means to parent. One woman on the Foundation website said: "It's easy to think that this was so terrible that it couldn't possibly have happened, even in your own mind. Sometimes you think like that. But it's so comforting to see a picture of him and say, yeah, I am a mom and he is my son and he was here." The picture is evidence of the baby's identity and also implies the identity of the mother, who may not be able to perform her maternal identity elsewhere. Particularly for those who do not have other children, these photos are important identity markers of them as a parent. Siblings and grandparents may also be depicted in the postmortem photographs, so it is not just parental identity but the familial unit more broadly that can be represented in these images. The identity of the family is central to the identity work in these remembrance photos.

Occasionally these media practices entangle familial identities with religious and even political identities. Expectant parents may learn of potential life-threatening developments or deformities early in their prenatal care. Depending on the timing and state regulations, some parents will choose to terminate the pregnancy whereas others will not, despite knowing that their child will likely not live long, if at all. This was the case of Heather and Patrick Walker and their son, Grayson.

Sixteen weeks into the pregnancy, the Walkers learned that their son had a rare neural tube defect that would lead him to be born without parts of the brain and skull. For religious reasons, they decided to carry the baby to term. The Walkers also invited a photographer to the hospital to take photos of Grayson during his birth and the eight hours he lived. These photos were posted to Heather Walker's blog as well as her Facebook page. However, they were flagged on Facebook as graphic images and the Walkers were asked to remove them. This story was picked up by the pro-life movement and reported in several online outlets. Pro-life groups lauded the Walkers for their religious commitment to celebrating Grayson's life as a child of God and were outraged that Facebook was demanding a grieving mother to remove photos of her recently deceased child. Facebook soon apologized to the Walkers for asking them to remove the photos and offered their condolences. There was also a pro-life grassroots movement on Facebook and elsewhere on the web to actively share the photos. Heather Walker received many social media messages "telling [her] how much God has used Grayson's life to touch them and to work in their hearts and to make them appreciate the children they have."[16]

This example demonstrates the political and religious nature of postmortem infant photography. These images can raise the questions of when is a life a life? When is this fetus a baby? When does pregnancy loss become mourning of a child? These are really hard questions without clear answers. Our political and religious identities often become central to our parental and familial identities that are represented in remembrance photography. Some say that if sex was the taboo subject in the nineteenth century, then death was the taboo subject in the twentieth century.[17] In the social media context, not only do these situations raise religious and political questions for the parents but the circulation of these images and stories can become highly politicized in their recirculation. Beyond political and religious moves, within the medical community we see contemporary shifts toward more open discussion and engagement with recently deceased loved ones.[18]

The Now I Lay Me Down To Sleep Foundation depoliticizes their work with testimonials of parents repeating the fact that everything was going fine in the pregnancy, suggesting there was no opportunity or need to contemplate pregnancy termination. Additionally, the foundation will only give parents the black and white or sepia-tone images that have been digitally retouched to correct for skin discoloration. Photographers are not allowed to give parents the original untouched color photographs from the hospital. The only images the parents get are beautiful images, often highly stylized, and "heirloom quality." Presumably the restrictions on sharing untouched images are to preserve the image of the foundation's high-quality portraiture. Moreover, it's not uncommon for professional photographers with the foundation to include a video of the collection of still images put to music with text overlaid, which can be shown at a memorial service.

Beyond the work of the foundation, there are hundreds of remembrance videos posted to YouTube celebrating the lives of both old and young. For most of these videos, postmortem photographs are not included because there are so many other media traces of the person from when they were alive and happy. But, in the case of prenatal or perinatal deaths, postmortem images can be found in these videos. And these inspire social remembrancing in the form of likes, comments, and condolences. In fact, most of these remembrance videos are made up of still images put to music. A common song is for these infant remembrance videos is Beyonce Knowles's song, "Halo":

Everywhere I'm looking now
I'm surrounded by your embrace
Baby, I can see your halo
You know you're my saving grace
You're everything I need and more
It's written all over your face
Baby, I can feel your halo
Pray it won't fade away

The song is clearly copyrighted but, unlike other YouTube videos, these videos are not taken down. The use of the song for infant memorial videos likely falls under fair use, where the purpose and character of the videos are clearly not for commercial gain.

Postmortem photos are still the central image of these videos because there are no other media traces to choose from. Often chronicling the experience of parents, some videos begin with images of the joy and excitement of pregnancy or even delivery, the initial joy of seeing the baby, then the devastating pain of knowing the baby is dying or dead. Sometimes other family members beyond the parents are depicted grieving for the baby in these remembrance videos. Chaplains are also depicted in these images either baptizing the baby or praying with the families. And then there are images of the dead baby being held and kissed by the parents. Often images include up close images of the baby's "sleeping" face or just their tiny feet. Sometimes the image is of the hands only next to the parents' hands or even wedding bands, symbolizing familial bonds. These images tend to come at the end of the videos, as if mimicking the grief more broadly—beginning with joy, moving to anguish, and then finally coming to peace with the loss. These traces help in the grief process as a way to both "let go but still hold on."[19]

Some of the remembrance videos on YouTube have hundreds of thousands of views, suggesting their circulation may be much wider than the immediate family and community directly impacted. Comments to the videos suggest that other parents or bystanders are also touched by the death of the child. Sometimes they reflect on their own loss, but moreover they feel the need to share supportive sentiments with the family sometimes years after the original video was posted. The loss of children is particularly seen as a collective tragedy and the circulation of remembrance videos can reinforce our need and display of social connection and support.

As a practice, remembrancing is a way of engaging with the present, past, and future through the creation of media traces. Remembrancing is a social process whereby we create and engage with traces of ourselves and those around us. Births and deaths are collectively celebrated and mourned by families and their communities, often through media accounting. We share our remembrances of these events as part of a collective memory process. Particularly in times of difficulty, we use remembrancing to reinforce social bonds of the family, of the community, and sometimes of our religious affiliations. We can use media accounting to document and share both wonderful and tragic life events so that we can collectively celebrate and mourn.

Remembrancing the Past, Today

Media accounting is particularly characterized by a presentism norm—what is going on right now, in the present? And the platforms themselves encourage this with prompts like "What's happening?" or "What's on your mind?" One common way of remembrancing occurs is by situating events of the past in the present. What happened today in my or our history? This is a common social media post genre. For example, there are many Twitter accounts that post the diary entries from famous people in history on that date. For example, @TheGeneral is a Twitter account for George Washington that frequently posts excerpts from letters Washington wrote on particular dates in history or posts about historical events in Washington's life #OTD (on this date) in history. Some have posted the journals of a dead parent as a way of memorializing their presence.

Newspapers for a long time have included sections such as "On this date in history," which will report on various events and activities throughout history that occurred on the particular date. Typically, these aren't long, in-depth analyses of the historical events, but one-sentence, factual accounts of the events of the day. Signing of treatises or laws, inaugurations, elections, as well as the births and deaths of famous people, are frequent fodder for such remembrancing. Thereby presentism norms of media accounting are maintained by remembrancing events of the past, today.

This kind of remembrancing also occurs within the domestic sphere. Anniversaries, birthday, and various kinds of holidays are similar occasions for everyday people to create media traces that reflect on activities or events of the past, through the celebration or mourning of that day. While posts saying happy birthday or happy anniversary are common on social media today, sometimes people celebrate such occasions by posting about their own involvement on that day: "Fourteen years ago today I married the most loving and patient man," or "five years ago today Ruth came into this world and changed our lives forever." Birthdays and graduations are common times for parents to post pictures of their children on the day they were born or when they started school, celebrating the present by remembering their past.

People also create remembrances of mournful events. On the fifteenth anniversary of the September 11, 2001, attacks, many people shared their stories of that day on social media. For example, Robert Glasper, a jazz pianist, tweeted: "Remembering those lost on 9/11. I was living in Brooklyn getting dressed to go to the airport when 1st plane hit." Many people on Facebook and Instagram posted images of American flags or of downtown New York City where the World Trade Center was. Some of the photos showed the present World Trade Center Memorial, but some showed images from 2001 of the towers after they had been hit, many tagged with "Never Forget." Zelizer argues photographs can be powerful collective memory tools through their symbolic value as well as their truth value.[20] Commemorative 9/11 posts are both a symbol of remembrance and a record of people remembering the tragedies of that day. Some found the images from the morning of the attack problematic. Several blog posts argued that people "shouldn't post photos from 9/11 on social media" because such tragic photos may cause personal pain for those directly involved.[21] For some these images were their media traces of the world around them; for others, the images are photojournalism.

Social Media Remembrancing Features

Digital and social media has made the recirculation of old media traces common. With ongoing usage of social media services, there has been a proliferation of specialized apps or platform features that enable and indeed encourage active engagement with previous posts. These services encourage people to reengage with and even explore and repost their previous postings. Facebook, in particular, has actively embraced remembrancing and its technological evolution demonstrates this. In 2011, Timeline was introduced, a major technological redesign that enabled people to readily and actively engage with their and others' previous posts. Instead of one's profile page being reverse-chronologically designed with one's most recent posts at the start, the Timeline profile reversed that. Timeline profiles started with your birth date and showed various kinds of Facebook activity by day, month and year. Indeed, the design of Timeline as the orienting framework of the profile page encouraged Facebook users to look back at theirs and others' previous posts.

Facebook Memories and Year in Review are other features on Facebook that encourage users to look back at their previous posts. When a person first goes to Facebook on a particular day, at the top of their newsfeed they

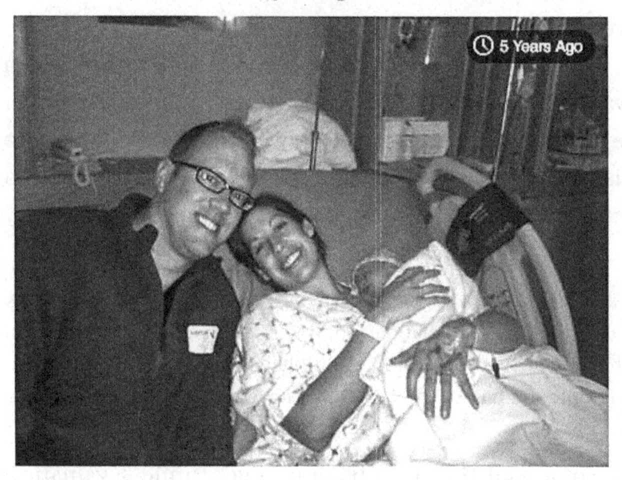

Figure 4.2
Facebook memory prompt: a screenshot of my Facebook Memories page, five years after our daughter was born.
Source: Author's personal collection.

may be prompted with a memory: "Your Memories on Facebook. We care about you and the memories you share here. We thought you'd like to look back on this post from five years ago" (see figure 4.2). Year in Review takes your Facebook posts from the previous year and algorithmically creates a video, featuring those events or memories that garnered the most likes and comments most prominently.

When Year in Review was first introduced, I watched the sample video of the Facebook employee who had traveled the world, gotten engaged, played frisbee, and hung out with friends, and was inspired to see my own memories beautifully curated for me. However, what I saw was a repetition of my daughter's Halloween costume. This post had garnered the most likes of any other of my posts because what two-year-old with big blue eyes doesn't look adorable in a fluffy duck costume? But this post was not my most important post. Not only is Halloween not a very important holiday to me, but I was surprised that *other* people featured so prominently in *my* Year in Review. I didn't have selfies of just me—they were always of me with someone. My Year in Review was more about the people around me than just me. We are social beings and thus share not only what we do but also what those we love do and accomplish as well.

For many others, however, the Year in Review and Memory features can be far more awkward or even traumatic. For example, in 2014 Eric Meyer blogged about the jarring pain he felt when he went on Facebook only to see a picture of his recently deceased daughter, Rebecca, with the caption "Eric, see what your year looked like!" As a designer, Meyer recognized why this occurred.

> To show me Rebecca's face and say "Here's what your year looked like!" is jarring. It feels wrong, and coming from an actual person, it would be wrong. Coming from code, it's just unfortunate. These are hard, hard problems. It isn't easy to programmatically figure out if a picture has a ton of Likes because it's hilarious, astounding, or heartbreaking.[22]

Tarleton Gillespie would call this a problem of algorithmic "evaluation of relevance."[23] There is no objective measure of relevance when deciding what kind of digital information to serve up to users. At best, Gillespie argues, engineers can approximate what seems to be or what is likely to be relevant. So for the majority of Facebook users, those faces or pictures that most often posted or commented on are likely to be representative of their year.

Facebook did not necessarily develop Facebook Memories and Year in Review because of a large sense of nostalgia nor to punish people who had lived through difficult experiences. Likely Facebook saw these features as a way to enhance user engagement with the site. Prompting previous posts can be a way to engage users who may not be producing content. Sharing these memories can be easier than creating something new to post. However, Facebook received a fair amount of public criticism for their Year in Review and Memories. Indeed, Facebook memory memes emerged in 2015, for example, to mock the experience of seeing happy photos of an ex-wife or ex-husband. In response, Facebook gave people options for managing their memories: users could choose to not receive 1) any memory notifications, 2) memories with certain people tagged, and 3) memories from particular dates or timespans.

Such solutions are problematic for several reasons. First, people likely won't set these restrictions until after they have already seen a troubling memory objects. Second, lots of people in photos are not tagged at all. For example, you cannot tag someone who's not on Facebook, such as young children or maybe elderly relatives. Third, dates of posts are not necessarily the date of photo. I may post a photo from days or months or years ago. It is not uncommon for people to take photos of printed photos to enable easier social media sharing, particularly with mobile devices. Restricting dates and Facebook users thus will not solve the problem because what triggers emotional reactions may not be part of the metadata attributed any one post, that is, when the photo was posted or who is tagged in it. Moreover, memories are not static but change over time. Facebook cannot account for this. Facebook only has access to memory objects in the form of posts. If someone has died, divorced, or lost their job, automated posts that remind people of this loss can feel cruel.

Summary

Remembrancing is a longstanding practice within media accounting. Whether it be creating or engaging with traces of difficult experiences, such as the loss of a child or sharing photos of the first day of school to commemorate the rite of passage, we use media accounting to connect with our pasts, presents, and futures. We anticipate memory work through the creation of mediated memories. We look forward, knowing that we will want or need to look back at various events and experiences through our media traces. Media accounting becomes anticipatory of wanting or sometimes needing to look back. The qualified self is shaped by looking back at traces of who we used to be, what we've done, and what we've experienced.

Remembrancing is also a practice of media accounting created with and intended for other people. When I take a picture of my daughter on her first day of kindergarten, I suspect that some day she will look back at that picture. She might not remember that day, but that photo will help her remember her teacher, her school, maybe a favorite pair of sneakers or backpack.

Maybe my son will look at that photo as well, as might my daughter's future children. While that photo felt personal to me when I created it, the sociality of remembrancing is central to the subject as well as future circulation of our media accounting. Thus, our qualified selves are entangled in the collective nature of our traces and mediated memory work. Social media make it especially easy to tag people as subjects in photos as well as to post it to their timeline, furthering socially entangling our remembrancing practices.

Remembrance photos taken of families who have lost a child are also collectively created and sometimes shared as well. While highly personal, this practice of creating and sharing remembrance photos reveals the complicated tension between public and private, and ephemerality and permanence. These images can help grieving families both hold on and let go. Media traces of loved ones who have passed are commonly shared at particular points in time, such as around a memorial service or funeral; however, the images' networked presence extends the future contexts in which such images may be found, shared, or circulated. Social media platforms which serve up memories may feel cruel to those who have divorced or experienced other forms of loss. The recirculation of tragic images in new contexts can separate the initial informational transmission purpose of the image with its recirculation. As such, images that were shared at a particular point in time for a particular purpose of mourning and grief may become startlingly graphic and distasteful outside the original context. Yet some memorial videos are watched hundreds of thousands of times by people not known to the family. Remembrancing the loss of a child, even if they are unknown to us, can touch us in emotionally powerful ways.

The presentism of media accounting coupled with our tendency to commemorate the anniversaries of various events, both joyous and mournful, continue to provide fodder for social media. These platforms increasingly leverage these tendencies to engage customers, serving up previous content in ways that may garner loving nostalgia or traumatic pain. Remembrancing is a way of creating, sharing, and engaging with media traces in ways that allow us to collectively think about our pasts, presents, and futures.

5 Reckoning

The fourth key practice of media accounting is reckoning. Reckoning is not a term frequently used in the United States, unlike other parts of the world like Australia or Great Britain, so I want to be explicit about how I use the term as a media accounting practice. Formally, *reckon* means to tell or describe, but it can also mean to count or measure out. Colloquially, *reckon* can mean to consider, understand, or think. Reckoning also suggests an explanation or evaluation. For example, to reckon financial accounts means to balance them. The act of describing or counting allows us to compare and evaluate. Media accounting thus enables us to understand, consider, and evaluate people, experiences, and events through the media traces created about them. When we see ourselves in media traces, we can scrutinize the traces differently than we can in our lived experience. Similarly, when we see media traces of others, we consider and understand people through their media traces.

Reckoning allows us to see things about ourselves and others that we may not be able to in our own lived experiences. Media accounting is not just the recording of activities or experiences, but is fundamentally a reflexive process that can reveal aspects or characteristics of lived events. Reckoning through media accounting allows us and others, whether they be family and friends or others, to understand ourselves and the world around us.

Reckoning can be considered what Joshua Meyrowitz calls self-surveillance, or the ability to record ourselves or have others record us in order to watch the recording in another time or place.[1] His quintessential example of this is the home video of a wedding. We might go to a wedding and experience it as a beautiful and lovely event. However, when we see the home video of it, we see sweaty people, awkward dancing, people talking with food in their mouths, and the inevitable drunk relative. We can scrutinize our media traces more easily than we do our lived experiences.

Through reckoning, media accounting becomes what Foucault would call technologies of the self. As media scholars like Jill Walker Rettberg have argued, social media are kinds of technologies of the self.[2] They enable self-disciplining, whereby we use media to better know ourselves so as to "improve" ourselves toward more normative expectations or ideals. Media accounting through diaries have long been considered technologies of the self.[3]

Historically, religious diaries were thought to help people reflect on their spirituality and in turn foster piety.[4] Writing in a journal about how one was or was not following the word of God externalizes one's behaviors and thoughts, which could then be scrutinized by oneself or others. Victorian parents who read their children's diaries aloud in the evenings were seeking to encourage not necessarily pious but proper behavior in their children.[5] Publicly sharing within the family what the children wrote they had been doing became an opportunity to reinforce positive behaviors and point out negative ones. In these cases, the subject of the diary was often the self, yet the social and collective values of the day are reflected in media accounting practices. The act of writing, and that which is written, becomes the reflection of both the self and society.

In the twentieth century, amateur film was used to better understand people and their behaviors. In particular, home movies became important sites of information about childhood. For example, in 1964 *Science* published an article about the use of home movies in psychoanalysis,[6] where media traces about family trips and holidays were used in conjunction with clinical practice to help patients better understand their own identity and relations. In the 1990s, home movies were used to evaluate early signs of autism.[7] In sports, video is used to help athletes see their behaviors and improve their technique.[8] The act of seeing yourself through media traces allows you to reckon or identify and understand trends that you might not in your lived experience. As Walker Rettberg argues, we have long understood ourselves through various forms of technology.

The practice of reckoning is fundamentally based on three particular aspects of media accounting: First is the evidentiary nature of media accounting. Second is the aggregation of information that is constitutive of media accounting. Third is the reconciliation process that is often tacit but central to the practice of media accounting.

Evidentiary Nature of Media Accounting

To say that media accountings are evidentiary is to argue that there is a veracity to the practice. That is, we understand and experience media accounting as evidence of actual events, activities, behaviors, or experiences. In photography, this characteristic is referred to as indexicality.[9] The easiest way to explain the powerful indexical nature of photography is the example of cheating in marriage. There is a big difference between seeing an illustration of your partner cheating on you and a photograph of your partner cheating on you. The indexical nature of a photograph indicates that the content of the photo is based directly on that scene actually happening before the camera, whereas the illustration does not have a direct tie to real events. But of course we know that photographs are not purely objective. Indeed, despite their indexical quality photographs can be subjective in their framing and timing. As such, photographs have a syntactic indeterminacy, that is, they can be interpreted multiple ways.[10] Therefore, there can be a difference between what is objectively observed in a photograph and what is subjectively inferred or interpreted from that image. The indexicality of images coupled with the syntactic indeterminacy suggest that they are both evidentiary as well as subjective.

We often interpret media accounting as indexical—something really happened and is known to us because of media accounting. This is certainly the case with the photos that are posted on social media, but even text within media accounting is interpreted as truthful. If someone wrote in their diary that it rained, then it likely did. It might not have rained all day or very much, but we presume that at some point it rained if someone wrote that it did. Media accountings—much like other forms of communication—are understood through certain logics or assumptions. In particular, we tend to presume a quality to our communication. The logic of quality, according to Grice, presumes a truthfulness of communication: "Try to make your contribution one that is true… Do not say what you believe to be false. Do not say that for which you lack adequate evidence."[11] We assume this truthfulness in our communication and therefore we use these logics as we engage in media accounting. There is a presumed veracity to all media accounting.

Therefore, when people create media traces of themselves and the world around them, the traces become evidence of events and experiences. There is a logical presumption of truthfulness when we read media traces. We also presume a kind of truth, veracity, or accuracy when we create media accounts. We typically don't make things up or purposefully lie in our accounts. That's not to say we don't imagine, pretend, and make believe through media accounting. Sometimes the hopes and dreams and desires that we put forth in our media accounting are actually a more accurate picture of who we are and the world we live in rather than what we might have done that day.

Media accounts like individuals' diaries and journals have long been used as historical evidence of economic, political, and social events and experiences of their day.[12] Similarly, we see that social media accounts are increasingly used as evidence of offline behavior and attributes. Social media traces are read as evidence of one's character that can be used in determining potential romantic partners or judging potential job candidates.[13] Tweets have been used in court cases as evidence of one's behaviors and intentions.[14] Family photos can provide visual evidence of a happy family.[15] A selfie can prove that we really did meet a celebrity or that we hiked to the top of a mountain.

Reckoning through media accounting is not just about understanding ourselves or other individuals, but understanding cultures, social values, and historical events. All media accountings have an evidentiary nature, which is what makes them so powerful to read or consume over time. We have evidence that something happened 150 years ago by reading someone's diary. Diaries, letters, and scrapbooks are preserved in museums and archives because they are evidence of the world long passed. A baby book is evidence that our fingers and toes really were once that small. Public tweets are being archived in the US Library of Congress because they are evidence of what is happening in the world.[16] Thus, media accounting is important to collective and personal reckoning because it reveals trends not only in individual traces but also in the aggregation of traces.

Evidentiary Nature of GoPro

GoPro is a camera company that was founded in 2001 by Nicholas Woodman. "GoPro helps people capture and share their lives' most meaningful experiences with others—to celebrate them together." GoPro produces action cameras that look like hearty two-by-three-inch black boxes and allow everyday people to produce professional quality videos to share with others. Indeed, GoPro has been quite successful at building and fostering a strong online community. In 2014, GoPro had an initial public offering to much fanfare.[17] The *New York Times* reported that "GoPro has evolved from exclusively building wearable cameras to focusing more on distributing the media that its cameras create."[18] While GoPro is fundamentally a camera company, social media, particularly YouTube, have been central to their success as a company.

What began as an idea to help athletes self-document while engaged in their sports, GoPro has become a standard for how people capture themselves engaged in their interests, whatever they may be. From extreme to mainstream, professional to consumer, GoPro aims to enable the world to capture and share its passions in the form of immersive and engaging content.[19]

The GoPro community is primarily made up of outdoor sports enthusiasts, like surfers, snowboarders, and skydivers who strap GoPro cameras to their heads or boards and use them to document their adventures. These athletes and enthusiasts not only capture their first-person perspective engaging in activities that most of us will never attempt, but they also provide evidence that they themselves accomplished these feats. These videos allow athletes and adventurers to both prove and improve their activities. By recording themselves they can prove that they accomplished various physical challenges, but can also improve their technique through closer scrutiny of their recorded behaviors.

Unlike most cameras, many GoPro cameras do not have a feedback screen or viewfinder to see what you are recording. The person recording cannot know what the camera sees until they are done recording. This feature presumes certain kinds of activities, like wearing the camera on your body. On the one hand, this aspect of the camera adds to its veracity because someone could not see how they were composing the shot. On the other hand, GoPro recordings are almost always planned because, unlike a cameraphone, GoPros are not typically carried on one's person. In fact, a "typical" GoPro video is often highly orchestrated both before recording begins as well as during editing.

A common kind of GoPro video is one by a surfer or snowboarder. Often this is from the first-person perspective; that is, with the camera strapped to a helmet or head, so we see what the adventurer sees. We can see the inside of the wave or the world upside down as they flip in the air over the snow, getting a perspective on the world that we may never get in our lived experience. Thus, we can begin to understand someone else's lived experience.

Increasingly, GoPro cameras are used to capture a variety of life's moments. From weddings, births, vacations to bike rides, middle-class Americans are using GoPro cameras (priced between $200 and $500 per camera) and ("free!") software to record and edit high-quality videos. For example, fourteen-year old Aldric Gozon strapped a GoPro Hero 4 to his chest and recorded himself performing Prelude no. 16 by F. Chopin on a Steigerman piano.[20] The video only reveals his arms in a dark long sleeve button-down shirt and hands furiously playing the keys. Behind the piano, we can see a beige wall and red curtain. In the reflection of the highly polished piano we can see what looks like a dining room table with chairs and the blinking red light of the GoPro camera.

We can see from his other YouTube videos that Gozon is a young pianist who competes in regional tournaments. This is the only one of his videos, however, to be recorded from his perspective. The other videos of his piano playing are obviously shot by another person in the audience of the competition. Yet, it was this video that was named a top GoPro video in 2016. Why? Because it captures much of what GoPro represents—perspective and accomplishment. From this perspective, rather than the audience view, we can see the dexterity and speed of his fingers moving across the piano keys. We can see what he sees. Beyond just distance, this perspective is far more intimate than the videos of his playing from the audience. There doesn't seem to be anyone else in the video, so it's as if he's playing for himself and that we are prying into a secret world.

Part of the allure of watching GoPro videos or reading diaries is that they provide a perspective on events, activities, and experiences that we may never get to experience. Diaries of famous men and women have long been sought-after because they provide an intimate perspective on events and experiences that we will never live through. Similarly, GoPro videos can provide perspectives on events, activities, and experiences that we may never live through. But we can come to consider and understand such experiences because we have read or seen the first-person accounts of others.

Sometimes we have experienced similar events and can learn from the media accounting of others. A surfer who watches a fellow's surfer's video can see subtle techniques in the video that might be lost on a nonsurfer. Similar, a farmer who watches a GoPro video of a first-cutting of hay sees things nonfarmers might not. The perspective and accomplishments of others read through their qualified selves can help us to reckon our own lived experiences.

Sometimes media accounting allow us to see the world we inhabit from a different perspective. Some of the featured GoPro videos are recorded from cameras strapped to dogs or even babies. The videos allow people to see the world they inhabit from a different perspective. This is reckoning—understanding the world around us through the media traces we create or that are created by others.

Sometimes the evidentiary nature of traces means changing our understanding of our lived personal experience. For example, GoPro videos can be valuable for the adventurer because they can see what worked or what didn't work. The trace becomes a means of self-surveillance.[21] The ability to record oneself can lead to the scrutiny of mundane behaviors, which can fundamentally change one's understanding of that behavior or event. The recorded trace has power over the lived experience because exposure to the recorded trace can replace or alter one's understanding of the event based on one's lived experience of it.[22] Therefore, power implicitly functions within Meyrowitz's concept of self-surveillance insomuch as media accounting allows users to "see" things about their behaviors they previously may not have perceived, thus changing their understanding of their own tendencies and behavior.

It is not uncommon for endorsed adventurers to use two or more GoPro cameras—one camera on their board or selfie stick recording the experience and one strapped to their helmet or chest to record what they see. The camera recording the experience facilitates self-surveillance so that they can understand and improve on their technique. But also records the other camera strapped to the person, which provides further evidence of the veracity of the final edited video as a media trace. Sometimes the first-person perspective is disorienting and the video of the actor helps to establish context. This is also accomplished with multiple adventurers wearing cameras, capturing the first-person perspective as well as broader contextual information with which to interpret the video.

Evidentiary Nature of Selfies

Selfies are the prototypical mobile media practice that leads to much social consternation and fretting. Selfies can be defined as a photographic practice of self-portraiture with the conflation of photographer and subject.

Concerns about narcissism frequently circulate around discussion of selfies, but there are various kinds of selfie practices.[23] Some selfies are individuals taking sexy photos to see evidence of themselves from different perspectives and to explore their sexuality.[24] Rather than understanding these practices of self-representation as narcissistic, these images can be seen as empowering. Couldry describes this practice as presencing; that is, the use of media by individuals, groups, and institutions to create a public presence beyond their bodily presence.[25] Richardson and Wilken offer a slightly more mobile-centric version of presencing, which is relevant to a discussion of selfies since selfies are almost always taken with mobile devices. Richardson and Wilken suggest that "mobile use crosses a spectrum of 'placing' and 'presencing.'"[26] Here the body-technology-place relations are phenomenologically contained within the device, within our situated encounters with the device, as well as within the broader networked environment.

In practice, some selfies are about presencing in a very literal sense, that is, providing visual evidence that I am or was somewhere. For example, Brazilian youth in disadvantaged urban communities would use selfies to let their parents know they were somewhere safe.[27] James Meese and his colleagues argue that presencing is an important function for selfies at funerals, suggesting that this seemingly disrespectful act is actually an important way that people engage in contemporary social rituals of loss. They argue that presencing is more than just positioning a subject in a particular context; it "immediately bring[s] that position to a wider social network."[28] Thus, the evidentiary nature of selfies at funerals allows people to bear witness socially, both through their physical presence at the funeral itself and through their broader network of social connections on various social media platforms.

Selfies with celebrities can also be understood also a form of presencing. This is not to be confused with celebrity selfies, or the popularity of celebrities like Kim Kardashian taking selfies and sharing them on social media. Selfies *with* celebrities is where seemingly everyday citizens take pictures of themselves with celebrities they meet on the street or in various public places. This selfie serves as evidence of co-presence with the celebrity. As I argued in chapter three, the term "selfie" becomes a misnomer because it is not just the photographer as subject, but as co-subject with the celebrity as evidence of the encounter.

Selfies are also commonly seen as before and after photos, as evidence of an "effective" weight loss or exercise regimens or even easier physical transformations like haircuts. Despite the evidentiary nature of media accounting, we often need context to reckon the media accounting of others. This can come from knowing the person, being with the person or having experienced the activities or events recorded. But sometimes it comes from multiple traces, like the example of before and after selfies.

Aggregation of Traces

The second key aspect of reckoning through media accounts is the aggregation of information, typically over time; as the GoPro example suggests, however, this is not always the case. Multiple GoPro cameras provide multiple perspectives through which to reckon an experience or event. Similarly, the diary, the album, the journal, the slide carousel, the YouTube channel, and the social media profile page are all collections of media traces that constitute our qualified selves. They are all platforms for media accounting practices. The collections of posts, bits and scraps are incredibly valuable. Indeed, the whole is greater than the sum of its parts. For it is in the aggregation that we can see trends, themes, or changes that we might not be able to in singular traces or even in our own lived experiences.

There is no set amount of information that constitutes media accounting, but it does tend to be more than one post, photo, or entry. It is the series or collection of media traces that constitute much of the value of media accounting. For example, Rettberg describes the power of serial selfies to cumulate self-presentations.[29] Sometimes as little as two traces can tell a very important story of media accounting. Before and after pictures are a perfect example of the value of media trace aggregation. It is in the juxtaposition of before and after that we can see the degree of change. We can see the growth of children over time. We can see how families or fashions change.

One of the key aspects of the aggregation of information with media accounting is that it is always incomplete.[30] The aggregation and simultaneous incompleteness of media accounting therefore means reckoning is about inferring, considering, weighing, and evaluating incomplete media traces. It is about interpreting and making sense of what happened in between posts or, to use Jacques Derrida's term, to understand the absent trace or "non-trace."[31] Simply put, reckoning is making sense of what is there as well as what is not there in our media accounting.

The Aggregation of Family Photos

The aggregation of media traces provides additional value and insight than would any one trace alone. Photo albums are common repositories for collective photographic traces, and the metaphor of the album has translated to social media. Photo frames can also aggregate media traces as well. Multipicture frames such as "My School Years" (see figure 5.1) are examples of aggregated media traces that reveal qualities or characteristics of a

Figure 5.1
"My School Years" multipicture frame made by Lawrence Frames. Printed with permission from Lawrence Frames, Bay Shore, New York.

developing child that might not be perceivable in lived experience in real time. This frame holds twelve little (wallet-size) pictures surrounding one 4x6 inch photo—one little picture from each grade in school—first grade through twelfth grade, culminating in a large picture of a graduating senior.

While you might look at this image and see bad '80s hair, I see the aggregation of information over time. If you can get past the hair, you can see the girl in the photos has a great smile. And despite the one change in sixth grade (an awkward time for many of us), she has had a consistent hairstyle. But the qualified selves reflected in this frame is not just the girl; the traces also represent her family, likely her mother. In this frame of photos, we also see the discipline of the parent who remembered to purchase and store the photos over twelve years, and we can see the care and love of the parent who brushed her hair and got her ready for "picture day" at school for many of those earlier years. The qualified self is not only in the media accounting of ourselves but also is reflected in the media traces we create of others. Reckoning is not just about the subject of media accounting; in this case, the girl in the photos. But reckoning can reveal insights into the creator of the media traces and the larger social context in which they live.

Today, most of the media traces we create are on Facebook or Instagram. And still we find great value and entertainment in the aggregation of media traces over time. We can see change as well as continuity over time. Both can be valuable and insightful. A common trend among new parents who are social media savvy is to take a photo each month during the first year of the baby's life. Babies are posed in the same location each month with some kind of sign with a number one through twelve to indicate how many months old the baby is. At the first birthday, the twelve images are then aggregated to show the changes in the baby. The aggregation of images reveals the emerging personality of this small person. But it also reveals much more. It serves as evidence of dedicated parents and their familial bonds with their babies. It reveals the dedication and devotion of the parent who, despite sleep deprivation, remembered each month to pose the baby in the same place. These carefully created traces of the child are "for" the child but also for the devoted parents themselves. The traces are evidence of a "good" parent and reinforce their own parental identities. These traces are also for the friends and family who will like, love, and comment on them. Some people will come to know these children primarily through the social media traces their parents create of the kids rather than through their own face-to-face interactions with them.

Reconciliation

The aggregation of information through media accounting directly leads to its third key aspect—reconciliation. As we engage with the media traces that we and others create, we reckon them with our understandings of ourselves and the world around us. When the traces do not match our understanding of events, some kind of reconciliation must occur. The reconciliation of media traces is part of the reckoning process.

Sometimes media traces confirm or support our understanding of ourselves and little reconciliation is needed. Other times, media traces counter our understandings or perspectives of ourselves and force us to reconcile our understandings. For example, when we see an unflattering photo of ourselves or our children, we typically do not reckon ourselves or them to be ugly, instead we reconcile it as a "bad picture." When we read excerpts from our diaries or old tweets we must reconcile those traces with who we are today. Is that still "us" or have we changed? Interpreting our previous media traces is fundamentally about reconciling our sense of selves with our media traces. When we read media accounting of others we also reconcile the traces with who we know them to be.

Reconciliation is sometimes about persuasion. Often media accounting is a way to persuasively and strategically change behaviors. Keeping track of what we eat or spend can help us eat or spend less or better. This is the basis of much persuasive technology[32] and central to technologies of the self.[33] When people are more aware of what they are doing, they can then change what they are doing. This is central to the quantified self-tracking movement.[34]

Like persuasion, reconciliation presumes the evidentiary nature of traces along with their aggregation but more importantly it is predicated on the incompleteness of media traces. The reconciliation of traces is fundamentally about the absent traces:

> Derrida's thesis is that presence is always already mediated by the absent trace; thus self consciousness is not a direct and unmediated experience but rather an indirect and always already mediated experience. This way of understanding personhood

and consciousness permits a key role for social and historical traces to enter and structure the very experience of consciousness and of self, even as those traces are unavailable to presence and awareness.[35]

Media traces can simultaneously reflect someone, their experiences or behaviors, but not reflect them at the same time. That bad photo of me is me, but it is not me. We reconcile it as a bad photo that caught me mid-blink because I don't typically walk around with my eyes half-open. Thus, the trace is true, but the absent trace is also true.

We are always in the process of making and remaking our qualified selves. Therefore, the media traces we create are not static representations of ourselves and the world around us, but rather should be understood as part of a reflective process. The reconciling that occurs is the ongoing interchange between the trace and the absent trace. It is the "always already" nature of the trace.[36] We anticipate the trace in our actions and words and thoughts, but the trace then reinforces and shapes our actions and words and thoughts.

Reconciliation of (Social) Media Traces

As part of my research I have studied several location-based social media services, like Foursquare and Dodgeball. These services allow users to "check in" at a venue and then share that information with people in their network. These services will also create aggregated maps of the places where users have checked in as well as others' check-ins within one's social network. For example, this is a map of my check-ins in Madison, Wisconsin (see figure 5.2). The black pins are where I checked in. The white pin is where my friend checked in and the gray pin is where we both checked in together.

Checking in on Dodgeball also enabled the social cataloging of people's lives. Dodgeball kept a history of where people have checked in, which could be plotted on a Google map, as well as imported into Google Calendar. For some people I interviewed, the mapping features became a motivating reason for their check-in practices. They liked having their "history" of where they had been cataloged in the system. Several informants mentioned that having this history of where they have been is "really fun" and that checking in to Dodgeball can be a passive way of documenting one's life.

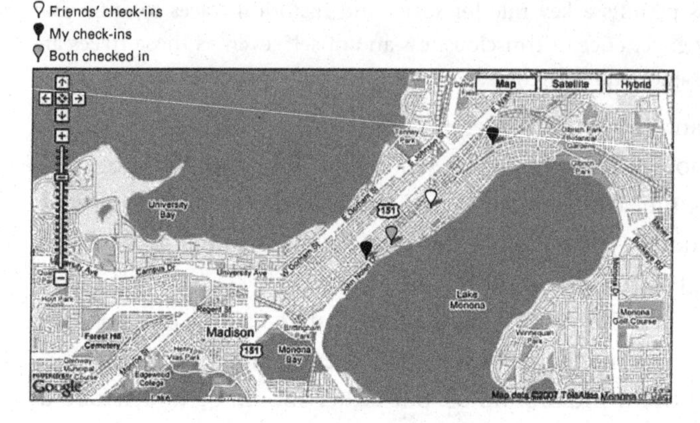

Figure 5.2
A map of my Dodgeball check-ins in Madison, Wisconsin, March 2, 2007.
Source: Author's personal collection.

I'm one of those people that tends to document everything that I do because it amuses me and because ... that's part of how I keep in touch with my friends. And I also do it because I have a really poor memory. So I've gotten in the habit of like documenting things on the Internet just so I have a memory there. And [Dodgeball has] really sort of enhanced that because it's more real time. You're not just writing about it after the fact. (Elicia, Minneapolis)

Elicia suggests that the real-time nature of the check in enhances the veracity of the trace. Unlike putting it into your calendar after the fact, by checking in at the location on the day and time, she felt the trace is better, more real, more accurate. By checking in at various locations, users have a record, a trace, a map of where they go and where they've been.

One participant I interviewed in New York City, Nick, pulled up the map of his check-ins. And as he reflected on it, he conveyed a sense of alarm that "according to the map," he had checked in several times on the Upper West Side of Manhattan, but he was a "Lower East Side" kind of guy, despite the fact that he actually lived in Brooklyn and worked in Chelsea. For those unfamiliar with New York City, the Upper West side is a fairly affluent residential area of the city, whereas the Lower East side is a more hipster neighborhood with an active nightlife and music scene. The map of his Dodgeball check-ins therefore did not represent where he "really" went and, more importantly, who he "really" was. He tried to explain that he recently had been hanging out with a friend who moved temporarily to the

Upper West Side, but that when he "goes out," that is, when he goes out to bars and clubs, he goes to the Lower East Side. Fundamentally, the traces did not match his sense of self.

Tensions that arise between our lived experience and the digital traces of our experiences reveal important questions of selectivity, veracity, and the qualified self. Was Nick's understanding of his social identity as represented by the venues he frequents wrong or was the map wrong? Perhaps the map was inaccurate because of a technical glitch. Or perhaps the map was accurate in what check-ins it portrayed, but it can only portray that information which it is fed. So perhaps just Nick forgot to check in to a bar or two. Or perhaps changing the timeframe of the map, showing check-ins from the past six months versus the last month, would have revealed more check-in activity in the Lower East Side. Or perhaps Dodgeball highlighted the activity on the Upper West Side. Such user activity is potentially more valuable to Dodgeball than activity on the Lower East Side given aggregated statistics about demographics or house pricing in the two neighborhoods (this is option is less likely). Regardless of the actual reason for the discrepancy between the map and Nick's understanding of his qualified self or at least that which he wanted to portray to me, he felt he had to explain to me why the traces were not aligned with who he "really" was. Therefore, Nick had to reconcile his qualified self with who he understands himself to be.

Sometimes reconciling media traces with our lived experiences means recognizing that things have changed. The recognition of change is often the valuable outcome of creating and engaging with media accounting. It is the reckoning that we have come from somewhere different than where we are today.

Mortified

Mortified is a project that is fundamentally about reconciling the media traces of our past with who we are today. *Mortified* was a grassroots stage production, podcast, and eventually a documentary film, two anthologies, and a TV series in which adults share media traces from their childhood. Sometimes these are photos or videos, but most often they are diary entries or poems that were written during difficult teenage years. Often highly humorous, these traces become ways to share sometimes very intimate details of our lives as a way to both connect with others and to connect with ourselves.

Self-forgiveness as well as humility are key elements of sharing embarrassing traces through *Mortified*. What makes these traces particularly embarrassing and often humorous is that they do not match who the authors are today, but that they are nevertheless traces of them. Describing a torrid teenage crush on a celebrity or the power struggles with parents over dating and curfews as an adult is embarrassing because those struggles no longer define who the authors are today. But those struggles previously defined them and in a very powerful way. Topics addressed by *Mortified* authors often include sex, family, friends, and love coupled with desperation, fear, anger, exhilaration, awkwardness, and desire. The *Mortified* documentary says, "The reason why it's so embarrassing and the reason why you're reading it is because it's still kind of you. Even though it's not you anymore and it's not that desperate. That's who you are and that's how you felt. And that is your truth." In this way, media accounting is about the reckoning that occurs as we re-meet ourselves through our media traces. *Mortified* is the very explicit and strategic means for people to explore their identities, their hopes and fears through their media traces.

For example, in one podcast, a woman describes her fights with her father over watching *The Love Boat*. As the woman describes the context in which she was writing about *The Love Boat* in her diary, we can see reconciliation. She informs the audience that as a teenager within one year three different adults in her life died tragically and that the fun and fantasy on the Love Boat was a way of escaping that difficult period. Her *Mortified* story was much bigger than *The Love Boat* or her diary entries about the fights with her father to watch it. Her story was about loss and hope. When we look back at our media traces, we bring to them perspective. What may seem trivial on the surface often reflects bigger themes in our lives.

Despite being personal reflections, *Mortified* stories are always social in nature. They recount other people in the author's lives—friends, family, crushes, celebrities. The relations, particularly familial, are often reinforced through the telling of the story and the reading of the traces. One young woman recounted her struggles with her parents through her diary, parents with whom she now has a very good relationship. But as a teenager and one of four children, she used her diary to assert anger and frustration with her parents in ways she couldn't or wouldn't to their faces. She swears and calls them horribly profane names in her diary, language with which time removes the ferociousness of emotion and seems funny in retrospect.

But *Mortified* is not just raw authenticity of human emotion and teenage struggle; it is produced. The reflections and media traces shared in *Mortified* are crafted by authors and producers of the show to create good stories. The documentary reveals some of this process as would-be *Mortified* authors are shown reading from their shoe-box of memories to producers who carefully help them curate their story. What would be of interest to an audience? A good story is immensely personal yet universal as the same time. As the audiences listen to *Mortified* authors, "people laugh or cry a little bit, but most of the time they identify with their own lives," the documentary tells us, because "we all went through the same pain and struggles."[38]

One of the key differences between *Mortified* and other popular story-based podcasts like *The Moth* is the importance of the media trace within the reconciliation process. For example, media become part of the performance as excerpts from diaries or letters are read and old photos are shown or described. These media traces serve several purposes. First, they provide evidence that these experiences really did happen. This is not just a story, but a *true* story. The veracity of a quote or a photo from the past is powerful in a media-saturated environment full of narratives both real and unreal that are obsessed with the present and future. Second, the traces become juxtaposed to the reflections of the authors in reckoning who they were and what they experienced with who they are today. That juxtaposition between me-then and me-now allows for the reckoning of me, my identity, and my social relations. Of course, this process is a recursive one, involving the creation of new media traces which are read in *Mortified* books, listened to on *Mortified* podcasts or watched in the *Mortified Nation* documentary film. These highly curated media traces are then shared and circulated online. The live show, however, does not rely on layering of media traces for effect. Instead, the liveness of the performance coupled with media traces from the past provide the juxtaposition of the author then and the author now. The live show, however, is featured prominently in the documentary film and in the social media traces about the shows. The layering and reconciliation of media traces allow for the reckoning process for both *Mortified* author and audience. As an audience member, listening to *Mortified* authors allows us to understand the author as well as ourselves.

Not only are the stories highly constructed to poignantly show how the authors' (universal) struggles in teenhood helped make them who they are today, but the performativity of sharing furthers the reconciliatory narrative of trace and selfhood. *Mortified* performers almost always mention other people in their lives, but despite the social nature of the content, sharing occurs on a stage with one lone *Mortified* author at a time under a single spot light reading from his or her diary or letters to the audience. Above them is projected a picture of them at the age when they created their media traces being shared. This sharing, or performativity of it, is central to reckoning through media accounting because sharing forces us to reconcile. Just as Nick felt obliged to justify his Upper West Side traces, the *Mortified* authors must also reconcile their traces because we are accountable for the traces we create or that are created about us.

The materiality of media accounting also features prominently in the performance. Authors clutch hard-bound journals in their hands from which they quote their former writings. From the documentary, however, you can see that these "diaries" also include typed-out entries, which have been taped or pasted into the journals, sometimes over handwritten text and sometimes onto blank pages. Presumably these are the reflections of authors as they help the audience and themselves to make sense of and reconcile their traces. Why did they write that in their journal? What did it mean? Where did it come from? What was the broader context in which it was written? These details are seldom included in historical media traces but are central to the reconciliation process. These are also highly produced by the *Mortified* author and the producers of the shows. The materiality of the paper or journals provide "evidence" of the performativity of media traces to further the authenticity of the *Mortified* moment of reconciliation.

As part of this media accounting process, each *Mortified* show ends with Lessons Learned. Each of the *Mortified* authors is brought back on stage as he or she is summarized with a statement about what lessons can be learned from each of their reconciliatory narratives of trace and selfhood. The culminating lesson learned from the aggregate *Mortified* readings is of course the tagline for the show: "We are freaks, we are fragile, and we all survived."

At a time when we are concerned about issues such as self-harm as well as bullying of young people, some have suggested that *Mortified* can be a way to inspire teen empathy. Listening to and sometimes laughing at adults who share their media traces may help young people to see that others have struggled with relationships, identity, and sexuality and that they are not alone in their own struggles.[39] "Share the shame" is another common way *Mortified* producers and authors describe the project. As part of the sharing process, presenters often reflect on the importance of sharing in helping to deal with difficult situations or experiences. But you can only "share the shame" once you have come to terms with what happened in the past and both reckoned and reconciled your media traces with who you are today.

#ThrowbackThursday

Reconciling media accounts occurs regularly on social media platforms like Facebook, Instagram, and Twitter. #ThrowbackThursday, or #TBT, is a social media convention that became popular in 2012. It typically involves people posting old photos of themselves. Childhood or baby photos as well as awkward teenage photos become fodder for social media interactions. While it would be easy to situate #TBT as a remembrancing practice, framing it as reconciliation provides a different lens for understanding this common media accounting practice.

Like Internet memes,[40] there are debates about qualifies within the boundaries of #TBT. For example, how old does a photo have to be to qualify as a #TBT post? Originally the trend involved photos or posts reflecting on years past; however, over time it certainly became more lenient. #TBT could be used from anything that happened in the past, even yesterday, reflecting the presentism of media accounting. Like all popular hashtags on social media, #TBT was also used by anyone who wanted to increase the popularity of their social media post. So if you searched for #TBT on Instagram, for example, you found many posts that are not of yesterday or years past, but instead use the hashtag to try to get more views and more likes to the post.

The popular press couched #TBT photos as fundamentally a nostalgic practice. In describing what the #TBT phenomenon is really about, "tech trends expert" Elise Moreau explained:

> People love to get nostalgic about their childhood, old friends and relationships, pop culture trends that are long gone, past trips or vacations and all sorts of other things that bring back happy memories. That's really all there is to it.[41]

#TBT was about looking back at our previous traces and fondly remembering past experiences. However, #TBT wasn't always about happy memories. Indeed, many photos that were shared revealed a very different version of ourselves than we might post regularly on social media. It is the difference between then and now that makes #TBT funny and socially compelling.

One etiquette guide in the mid-2010s written by Steven Petrow articulated exactly what #TBT posts should encapsulate: "Photos should be funny or really what I'd call 'self-funny,' even a bit narcissistic, which is to say you're mocking yourself and not someone else in the photo. Embarrassing is good. Or horribly, horribly cute (again, it needs to be you)."[42] As *Mashable* writer Brian Koerber wrote, "Who doesn't love a good photo blast from the past? There's nothing wrong with enjoying photos from your awkward years, like that embarrassing haircut you had in middle school."[43] According to these guides, #TBT photos are a means of sharing embarrassing or awkward photos from your past.

Like many etiquette guides, Petrow and Koerder make explicit social norms that are typically implicit. Etiquette guides for new media are particularly helpful to technology scholars because the norms are often still emerging, that is, new users of these technologies need guides on how early adopters of these technologies have defined "appropriate usage." These etiquette rules can relax over time, but their initial articulation points to a "proper" or "good" way of doing something, thus moralizing very mundane and everyday activities. "Good" here means that your usage will be received well by others. Etiquette guides to technology are no longer needed when the general public understands and follows the norms of technology use.

Other #TBT etiquette guides in the popular press articulated further aspects of #TBT: posting too many reminiscence posts per week is bad (one is plenty and preferably on Thursdays); photos should be of yourself but can include others, as long as you do not tag the other people without their permission; make it old (not recent); don't share photos of meeting a celebrity (no #humblebrag); and make sure it's a good photo of a photo.

This last point regarding the "photo of a photo" is an important aspect about #TBT. For many people who were young before the advent of networked photography, the only images they have of their childhood or teen years are printed photographs. Therefore, to post them to social media, people have to digitize them, either by scanning them or by taking a digital photo of the photo itself. The latter is by far the easier way to share a printed photo, however, the quality of the image is typically significantly lower. Sometimes the angle reveals that it's a photo of a photo and other times one can see the hand holding the photo or the table it is on. "Good"

#TBT images should not reveal that current world in which the image sits. Instead, the juxtaposition between then and now reveals itself in the aesthetics of the content of the image: in the dress, the haircuts, the poses of the people portrayed. It's in the color of the images. Tables and fingers draw our attention away from the subjects of images in ways that minimized the impact of the image. When a good #TBT image presents itself as any other social media post, the aesthetic contrasts and awkward poses become reconciled with our contemporary understandings of who we are in socially meaningful ways. The contrast between the presentism of social media traces ("this just happened") with the historical trace necessitates reconciliation. In particular, the hashtag #TBT itself serves to signal the reconciliation between then and now. #TBT shows that the poster knows they are violating the norms of presentism on social media.

Beyond signaling norms, #TBT is really about reconciling who we are now with who we were then, much like the *Mortified* project. We can laugh at the photos of faces we made, haircuts we had, and clothes that we wore in our past because we don't have them anymore. We've changed; we've grown older. We can laugh at ourselves, so that others can laugh with us.

Media accounting is about reconciling our sense of selves with the traces we create or have created in the past. The juxtaposition between then and now or between trace and experience must come to some equilibrium. Why is there a difference? Is it a bad picture or is it an old picture? Was there a mistake? We need to understand and make sense of our media traces. They need to align with our understanding of ourselves and our understanding of the world around us.

To reconcile means both to bring to peace or restore, but it also can mean to make consistent or compatible with each other. The qualified self often involves reconciling our traces with our understanding of our experiences and sense of self to make our traces compatible with our experience and to restore our understanding of who we are and who we were.

Summary

In this chapter, I detail the practice of reckoning through media accounting. Reckoning is the means through which we and others come to understand our qualified selves through the media traces we create or that others create about us. Media accounting has a presumed truthfulness to it which functions within the logics of communication. As such, written or visual traces we create also have an evidentiary nature to them. The pics, posts, and entries are all taken as true, as evidence of something that happened. Selfies or first-person action cams like GoPro can prove that we were somewhere, did something, or met someone. Media traces can be scrutinized in ways that lived experiences cannot. Thus, selfies and GoPros can offer us ways to see ourselves from a different perspective as well as to let others see us, our actions, and experiences.

The aggregated nature of media accounting provides valuable information about ourselves and our world, whereby we can see trends not discernible in our lived experiences. But there is a disciplining aspect to the aggregation of traces. It takes dedication and work to create, collect, and maintain traces over time. Indeed, those who would keep the most detailed of media accounts may not be the most ordinary of actors.[44] Media accounting is therefore defined by its incompleteness or discontinuity. Thus, the media traces we create are simultaneously also evidence of a non-trace—what we didn't capture or what we didn't create traces of. While diaries and photo albums are containers for aggregated traces that offer potential end points with finite pages, social media platforms are not. Thus, social media platforms highlight the continued incompleteness of media accounting. One is never done. There is always more to document and share.

Most of the time the media traces that we share confirm or affirm our sense of ourselves. But sometimes they don't, and we must reconcile the trace with the sense of self. Reconciliation is a social process whereby we bring our media traces and our understandings into equilibrium. Reconciliation can be seen in narratives about our media traces, and in the explanations and contextualizations that accompany our traces. This can be in an interview of a mobile social network user or a performance of sharing old diaries or even in a hashtag like #TBT. When our traces, which become aggregated evidence of who we are, are *not* who we are, we need to engage in the media accounting work of reconciliation. Our qualified selves are always being remade and reshaped in light of new and old media traces. Reconciliation is a common process through which we can realign our qualified selves.

Reckoning and reconciliation highlight several media accounting dialectics. First, the disciplining or persuasive aspects of self-tracking through media traces highlights tensions between work and leisure. Documenting our own physical changes or our children's development over time takes self-discipline in many ways, but may be personally and socially gratifying. Whether it be skydiving or playing the piano, the work of mastering physical accomplishments may be part of a leisure activity or hobby, but takes dedication and resources.

Reckoning also highlights the tension between ephemerality and permanence. The trace always already implies the non-trace, that which isn't captured. So, despite the evidentiary nature of media traces, questions regarding what happened before or after the trace are central to understanding the trace itself. The multiphoto frame becomes an aggregated representation of twelve years of living, not just for the subject of the photos but the people who helped to make those media traces happen. The frame both reflects and captures the fleeting nature of childhood. Change and understanding occur through the dialectic of ephemerality and permanence of our media accounting whereby the trace is actual and permanent but nevertheless also fleeting. The practice of reckoning through media accounting relies on the permanence of traces while simultaneously recognizing our inability to go back to that moment when the traces was created.

6 Conclusion

Throughout the previous four chapters, I have outlined the key media accounting practices at work both through diaries and photo albums as well as through contemporary social media platforms. I argue that an important use of media in our everyday lives is to document life and to share it with others. While often heralded as a unique characteristic of Web 2.0, people have long created or authored media for themselves and others. Media accountings are overlooked practices that reveal the ways we have incorporated media into our everyday lives for hundreds of years.

Each of the four practices reveals a different tension that social media have been accused of bringing about. Sharing quotidian aspects of our lives through media accounting is the first practice that characterizes this genre of media production and use. Using ritual theory, I demonstrate how the act of creating media traces of our experiences and of those around us transforms experiences into ritualized accounts that can reflect and reinforce the social order. Critiques that social media have blurred the public and private distinctions are revealed to be less helpful when thinking about journaling practices in the eighteenth and nineteenth centuries. Indeed people, particularly women and children, have long negotiated publicness and privateness in their journaling practices. That said, the degree of publicness is potentially greater in the networked environment, where videos and posts go viral,[1] but also where broadcast media increasingly report on "trending" topics online, further amplifying the publicness of any one post. More commonly, however, blogs draw on the diary genre in similar ways to those who publish their diaries while still alive. We have long enjoyed reading or watching the daily experiences and reflections of those willing to publicize it.

The relational aspects of the qualified self necessitate understanding the identity work of media accounting. The content of our media traces, even our selfies, are seldom about just us. Our relationships and interactions with others are represented in our media traces. Sometimes we create traces for others, like baby books, which still reflect our own identities. It's not just the content of our media traces but the circulation more broadly that reveals the individual and collective dialectic. This is particularly true for women and families, whose media accountings are collectively shared. The key difference in the contemporary social media environment versus that of the eighteenth century follows similar trends in twentieth-century domestic technology.[2] That is, increases in technological domestic capabilities led to increased expectation for domestic labor, particularly for women. In this case, it has led to rising expectations for familial media accounting, resulting in substantial work on the part of women and mothers. Ostensibly mobile and social media have made it easier to media account for the family. As I argue, mobile phones are the primary means of creation, distribution, and consumption of media accounting. But expectations for media accounting have risen. These changing expectations are true for men and women, but women are typically responsible for accounting for the family as well as being accountable to receive and recognize the media accounting of others.

Remembrancing highlights tensions of ephemerality and permanence in our lives. We have long documented both joyous and difficult moments; however, we can neither document everything nor just let it all go. While we may try to document our travels, we have to leave out details. Sometimes we don't want to create a trace of something, especially tragic events. However, only by creating ways to hold on to something really difficult, such as photographing the loss of a child, can we begin to let go and heal from our loss. The social media environment makes the continued engagement with previous traces both nostalgic and fun as well as potentially traumatic. We engage with the past through the present. Sometimes this is purposeful, such as when we commemorate and connect to the past by remembrancing what happened today in history. While people can gain social support in sharing their traces, the algorithmic recirculation of the traces removes their context, making some traces seem harsh or taboo. In other cases, the networked environment means media traces created in a interpersonal context become quickly taken up for broader political purposes.

Last, we come to understand who we are and how we have come to be through media accounting. While diaries, scrapbooks, photo albums, and social media profiles are always already incomplete, media accounting provides importance evidence of the processes of our experiences and relations. Reckoning occurs through our inferences of the aggregation of information over time, that is, through the compilation of our media traces. But sometimes the traces do not match our sense of selves and we must reconcile the trace with how we understand the world and our place in it. Much of coming to understand the qualified self is a reckoning process of subtle consideration. While literature on the role of technologies of the self and self-tracking often uses Foucault's work, reckoning the qualified self is a more ordinary and softer process of change, one that focuses on reconciliation rather than discipline.[3] Social media platforms are not only sites for the practice of reckoning, like other media accounting technologies, but also forums for public reconciliation of the qualified self. The sharing of old posts and photos encourages people to reflect on their media accounting and how they and their relations have changed, yet also remain the same.

While the practices of media accounting remain consistent, the modes through which we engage in these practices change over time. Those who engage in media accounting adapt to and adopt new technologies. Tracing changes in dominant modes of media accounting over time reveals differences not just in the contemporary social media environment but in how mediatization has impacted media accounting throughout history.

Mediatization and Media Accounting

Changes in media accounting over time can be understood through changes in mediatization. Nick Couldry and Andreas Hepp write that mediatization is a meta-process of social change whereby increasingly mediated communication is understood as a transcultural and dialectical processes of transformation at every level of interaction across society:

> Mediatization is a term that enables us to grasp how, over time, the consequences that multiple processes of mediated communication have for the construction of the social world have *themselves* changed with the emergence of different kinds of media and different types of relation between media.[4]

Couldry and Hepp argue there have been three overlapping waves of mediatization in history: mechanization, electrification, and digitalization.

The mechanization wave of mediatization fundamentally influenced the production of paper and journals (in addition to books) in the fifteenth century in Europe and had profound implications with regard to journaling.[5] While paper and journals are not considered mechanical, it is the mechanization of their production that led to a standardization of journals and journaling more broadly beyond their religious purposes.[6] The mechanization wave of mediatization, coupled with shifts from orality to literacy, set the stage for media accounting.[7] Journaling became a media platform for the aggregation of personal writing about everyday events not just by powerful leaders, explorers, or clergy but by everyday people. Journals first enabled the creation, collection, and sharing of representations through media traces, which are essential to the qualified self.

The electrification wave of mediatization transformed media accounting in two ways. First, photography as part of electrification offered new modes of media accounting. No longer did people have to document solely through writing or the occasional drawing; they could now photograph and film people and world around them. They could capture life events, including the deaths of their youngest family members. In the late nineteenth and early twentieth century, we see snapshots emerge as an important mode of media accounting that continued throughout the twentieth century. Second, electrification also had an overshadowing effect on the continued mechanized mediatization of media accounting. Electrification brought about new media organizations and social institutions, which contributed to great societal change. These were huge changes for globalization, communication, and society, but not necessarily for media accounting. Nick Couldry and Andreas Hepp explain their use of the wave metaphor to describe these phases of mediatization.[8] Waves have peaks and crests, but they are still connected to and part of the larger media environment. Older media do not disappear in light of new media. But a society focused on the peaks of telegraphy, radio, film, and television no longer paid attention to the larger ocean of media, which still included journals, diaries, and scrapbooks. During the wave of electrification, "the home mode of communication"—as Richard Chalfen calls it—went largely unnoticed by media scholars.[9]

The digitalization wave of mediatization ultimately changed the visibility and circulation patterns of media accounting. Digitalization is associated with the rise of the Internet and social networking platforms that enabled an increased visibility of media accounting. No longer just audiences, we began to recognize that people could be media producers too.[10] Of course, they always have been with regard to media accounting. However, with digitalization media accounting no longer circulated among interpersonal networks subtly under the electrification wave of mediatization. Digitalization brought an increased visibility of media accounting—most prominently through blogs. Cheaper, more integrated mobile technologies further enabled the creation, distribution, and consumption of media accounting. As a result, people collectively become more aware of engaging in media accounting because they increasingly received and engaged with others' qualified selves.

Searchability and replicability are considered further transformative characteristics of the digital environment.[11] The ability to search for media accounting of particular people or to search for content within media accounting led to greater visibility of media accounting. Replicability also meant increased sharing of media traces which further enhanced our collective visibility of media accounting.

As the interconnections between people, media technologies, and media institutions grew during the digitalization wave, some people became more avid documenters of their lives as well.[12] The ubiquity of camera phones changed expectations for visual representations of everyday life. Despite Kodak's high household market saturation in the 1980s, camera phones meant that not only did every member of the family have a camera but also they constantly carried it with them. Eventually notions of self-documentation become entangled with social media platforms themselves. Phrases like "if you didn't take a picture of it, it didn't happen" turned to "if it wasn't posted on Facebook, it didn't happen." You might have gotten married or given birth to your child, but until you created a trace of it on Facebook, it did not *really* happen. Until you create media traces and share them with others, events are not complete.[13]

Throughout the waves of mediatization there has also been an increased role of media institutions. The increase in media accounting visibility due to digitalization was not only among people, but media institutions increasingly gained access to our media accounting as well.

Media Institutions and Media Accounting

Comparisons between historical media accounting practices and media accounting on social media platforms today can be made in terms of media institutions' roles in media accounting. As chapter 3 argues, the commercial nature of media accounting, that is, the use of commercially available products to engage in media accounting, is not a new aspect of its practices. For over two hundred years, people have bought journals and, more recently, cameras, photo albums, and scrapbooks with which to document their lives and the world around them. While social media platforms may be "free" to download on Google Play and Apple's App Store, the media technologies through which we access these platforms are not. Today people primarily buy mobile phones through which they engage in media accounting. People must pay for phones, phone service, and increasingly phone repair. Some put cost estimates for American smartphone use as high as $1,000 per year per family.[14] And while we might think of social media as communication *services*, they are commercial products just like earlier media accounting platforms.[15] Commercial products have always been used for media accounting, but the role of the who *owns* these media accounting platforms requires further explication.

Media Accounting Ownership

Prior to contemporary mobile and social media platforms, once someone bought a journal, for example, they owned it and they owned the content they created within it. Or at least so it would seem. But when Victorian parents would read their children's diaries out loud, who owned those diaries? The child? The parents? The family? When husbands and wives would send journals through the mail both detailing their own experiences and commenting on the others' experiences, who owned those journals? When someone dies and their children or grandchildren find their journals, do they become the owners of the journals? Such historical family documents might circulate within the extended family. And, if we are lucky, such media accounting might find its way into an archive. While purchasing particular media accounting platforms might lead to individual ownership in some circumstances, the *circulation* practices of media accounting complicate simple delineations of ownership.

Additionally, as I have argued, the relational nature of the qualified self also complicates media accounting ownership. Other people figure into our media traces and sometimes traces are made on behalf of others. For example, who owns a baby book? It is typically made by a mother on behalf of her child. While the mother might maintain and retain the media traces throughout childhood, it is the child's book. Ownership of media accounting has never been simply individualistic, but is complicated through our interpersonal networks of media accounting production as well as their circulation and consumption.

But what about the role of the commercial entities who provide us these media accounting products or platforms? Indeed, there are ownership differences between the journal or diary producers and social media platforms today. Historically paper mills would sell journals to markets or stores who would sell them to customers. Upon buying them, neither the paper mill nor the store would have any claims over the journals nor their content. Today social media companies typically retain ownership of their product, that is, the platform, the app, the website, but users own their content. For example, on Instagram users retain ownership of the content they produce, that is, they retain copyright of their photos. But the terms of service to use the platform, like most social media platforms' terms of service, give Instagram license to use and share "user content." Here user content is not only our posts, pictures, comments, and hashtags but also our profile and network information as well as metadata about our usage. Legally, users own their social media content but many others have access and can use it.

But the fact that the company retains ownership of the product or platform through which we engage in media accounting does have implications. First, it means we cannot truly modify the platform itself. Anything that we change, such as the background or font size is from a preestablished set of options by the company itself. Despite their APIs, these platforms lack what Jonathan Zittrain calls the "generativity" that characterized much of the early Internet.[16] Historically, we can see such modifications, particularly in scrapbooking. People would not just reuse commercial paper, but would write their own content over the content produced by others. They would tear pages from books with which to make their own books and write on. In the social media environment today, we can only modify that which social media companies allow us to modify. Second, the fact that social media companies retain ownership of their products means that the products can change while we're using them. This ostensibly means that the companies can continue to improve their products without us having to buy new ones, but it can also be frustrating. As interface and design features change overnight, we must relearn the ways to engage in media accounting on specific platforms.[17] Social media companies' distinction between platform ownership and content ownership is a key contemporary development in media accounting.

Social media companies are not the first ones to retain ownership of their products. There are examples throughout the twentieth century of users not owning the media technologies they use everyday. For example, telecommunications companies frequently retained ownership of the products they serviced. Landline telephones were often leased from telephone companies.[18] In the United States today, the cable boxes, cable modems, and even television remotes in people's homes are also typically owned by the cable companies.

Media Accounting Access and Commodification

Distinctions in ownership between product and content now lead us to ask about *access* to the content itself. While paper and journal producers in the nineteenth century made money from selling these products, they did not own nor have access to the content of the journals they made, that is, the media accountings themselves. However, this is not true for all historical media accounting platforms. At the end of the nineteenth century, Kodak sold cameras with film. People could take up to 100 pictures, then take their camera and film back to Kodak for development and a new role of film.[19] In this case the company had access to the content that users were producing. Eventually the process of film development expanded beyond just that of Kodak, allowing independent film developers and processors to emerge. But for at least the early years, Kodak developed the film for their cameras and therefore had access to all of the content their customers produced or created through their technology. So, commercial access to media accounting is something that has occurred historically and not just with social media, but it looks different today.

Today, social media companies have complete access to the content of our media accountings on their platforms; however, we do more on social media than just document our lives and share it with others. We watch funny videos and read news articles. We make our social connections explicit through our "friends" and "followers." So the social media companies *also* have access to our network connections, and our networks' network connections. They have metadata regarding when we log on and for how long, from what device, and what we look at, click on, and "like" on the website or app. This is coupled with our profile data, including where we live, and sometimes also our education and for whom we work. They aggregate all of this data and come to understand their users and the system more broadly. Fundamentally this extensive access transforms ways of knowing, what is known, and what is knowable.[20] The extent of social media companies' knowledge about users is far beyond that which previous companies who served or produced media accounting platforms ever had.

A key difference therefore between historical media accounting platforms like diaries, scrapbooks, and photo albums, is what they do with this extensive user knowledge. Today social media companies commodify that information by selling advertisements.[21] While Kodak processed millions of customer photos, they did not share that information with advertisers in exchange for access to their customers. In other words, Kodak did not commodify its users. They made money by selling their products and services, not from advertising. The commercial model of social media companies fundamentally follows an American broadcast model of media, in which users pay for their media not with money but with their attention. Ultimately, however, social media users do not pay for media merely with their attention; their usage becomes the product that is sold to advertisers. This shift is paramount to Couldry and Hepp's potential fourth wave of mediatization: datafication.[22] But as I argued in the introduction, while datafication is relevant to media accounting, the qualified self is ultimately better understood through mediation, that is, the reciprocal relationship of influence between media and people. Therefore, the further mediatization of media accounting through datafication can be understood and countered with what we might call a postdigital turn.

Postdigital Media Accounting

The media studies scholar Sy Taffel suggests that the postdigital counters a variety of tensions that the digital has encompassed. Namely, the postdigital represents:

> 1) a return of the analogue or move beyond discrete samples, (2) the revelation of seams and artifices within the otherwise smooth spaces of the digital, (3) the historical phase of technocultural development occurring after the digital revolution, (4) the rematerializations of digital technology and its integration into urban environments and (5) a way of escaping the fetishization of newness and upgrade culture.[23]

The postdigital of media accounting as a return to analogue can be seen in the proliferation of old-fashioned-looking moleskin journals and the growing industry around journaling as crafting. As Jonathan Sterne points out, analogue is not just that which is not digital, but has come to represent a larger cultural condition despite its technical etymology.[24]

But the "post" in postdigital is not necessarily a rejection of or break from the digital. The "post" in postdigital can convey a continuation of it. Therefore, postdigital is that which comes *after* the digital.[25] As Taffel's fourth definition suggests, the postdigital can be the rematerialization of the digital. Within media accounting, the postdigital can also be understood as that which comes after our traces on social media through their rematerialization.

There has been a recent proliferation in products and services which allow us to create material forms of our digital traces. From books of digital photographs to wall art, calendars, cards, and even T-shirts, these products seek to take our digital media accounting and make them material. For example, Prynt is a company that creates small cases which smartphones snap into and enable printing on demand, turning any camera phone into a Polaroid camera. The Prynt promotional video depicts young men and women hanging out together, then one of the women snaps a picture of the group having fun and prints it.[26] Everyone laughs seeing themselves in the photo. Once printed, the photos are then used to decorate. One posts it on a fridge, one hangs it from a tree, and another peels off the back to reveal a sticker, which she playfully sticks to a young man's forehead. Only women or feminized hands are depicted in the video as taking and printing photos, suggesting that it's the role of women to document the social. While men might look on and enjoy, it is the work of women to carry the Prynt device with them, to decide what is "Prynt-able," and to take the time out of socializing to edit and print the image. The act of creating the image is depicted as great fun, but the creation of the printed photo itself becomes an additional object of mutual gaze and adornment.

Prynt's line of products fit into what Caroline Basset calls a "object-oriented ontology" (OOO), that is, the postdigital proliferation of sometimes weird material objects in response to the digital.[27] While Basset critiques much of the OOO within the art world as focusing on predominantly masculine objects, postdigital media accounting focuses on more feminized objects. Like the Prynt video suggests, women are likely to be the ones to document a social event and share it with others. Photo services like Snapfish and Shutterfly not only offer prints of digital photographs but can turn those images into books, posters, wooden prints, calendars and so forth. Even Apple recognizes the tendency to create material forms of digital traces and offers "Projects," which enable people to create cards, books, and calendars, as well as prints of their photos.

But postdigital media accounting moves beyond the printing of camera phone photos to capture our various social media traces—including captions, conversations, and hashtags. For example, My Social Book is a service that takes postings from Facebook and Instagram and creates a book: "Get the Book of your Life that will last Forever."[28] Customers can create books of someone's entire timeline or all posts with a particular friend. The company can create books of what they call "blended timelines," marketed as "My Social Book Lovers," which seems to take popular posts from when two people were not connected on Facebook and Instagram to when they become connected as friends or by their relationship status. My Social Book is predicated on the idea that people are posting their lives to Facebook.

Yet, a postdigital reading of My Social Book reveals three tensions of media accounting that arise on social media platforms. First, My Social Book becomes a means of reclaiming ownership of one's digital traces. The digitalization wave of media accounting brought about new levels of corporate visibility and access to our media accounting. Physically printing our traces is way to reclaim ownership of our media accounting from the platforms and technologies used to create them. We *own* our books and their content. We can now really own our social media accounting on the shelf or wall in our house. Second, postdigital media accounting enables people to physically hold on to their memories, *forever*. My Social Book is a keepsake book. Facebook is not a keepsake. It is presentist, it is ephemeral, it is in the cloud. My Social Book will last long after Facebook goes away, as a business and as a platform, or so they suggest. Third, the mutual exclusivity of material objects increases their gift value. It is much more meaningful to give someone a physical book or calendar with their and our photos in it than it is to email them with such a collection or to tag them in it. There is a presumed cost associated with printing such items but it also takes times to create such material objects. (Despite templates, I for one have spent hours upon hours deciding on and arranging photobooks for grandparents.) Postdigital media accounting solves regarding some tensions regarding ephemerality and ownership. However, postdigital media accounting creates further tensions regarding the amount of resources it takes to create these traces. The time and effort it takes to create postdigital media accountings has become factored into the planning of some of life's major events.

The Qualified Self Wedding

In 2013, the W Hotel in New York announced that it was offering a "concierge service" as part of its wedding packages. For a mere $3,000, the hotel would provide a custom hashtag for a wedding; track and collect all of the social media posts on Twitter, Facebook, and Instagram; and then create a Shutterfly hardcover book of all the images and posts related to the event. The book becomes the material instantiation of tweets and Instagrams and becomes an important keepsake to capture this momentous event. But of course the couple will still need the professional photographer and videographer to chronicle the event for them and provide the official narrative of the events. The increasing commodification of weddings necessitates the professionalization of media accounting of the event. The W's concierge service solves the problem that social media traces created by others around this event are seldom captured or done so in only an ad hoc post-facto manner. The W offers couples the ability to catalog through a "custom hashtag," but also provides the couple with a document of the traces. Like putting disposable cameras on tables and encouraging guests to document their versions of the events, the W is offering a fuller version of the couple's wedding—a version full of ritual and rite but also full of humorous pics and tweets which might not make it into the official wedding album or video.

Moreover, this kind of service both recognizes that people participating in the event are coproducers of it through their attendance but also through their own qualified self practices. Like all media accounting, media traces are seldom only about the author himself or herself, but instead the context in which someone creates a trace is socially determined. For example, a guest's selfie at my wedding is something that I want in my media accounting of the event. The qualified self wedding therefore becomes collectively made and represented in media traces of those in attendance but also those who view and engage with such traces online.

The W's "custom hashtag" service helps to collect the digital traces surrounding the wedding in the same way that guestbooks and disposable cameras on the reception tables register people's presence at the event. They become part of the collective sharing, identity work, remembrancing, and reckoning of the qualified self wedding. But now even those who are not in physical attendance can respond and contribute to the social media traces. The W's use of the term "custom," however, conveys the W's commodification not only of the event and its media accounting but of hashtags themselves. The outsourcing of media accounting is not new. Ellen Gruber Garvey found that scrapbooking services emerged which would make beautiful books of news clippings, but it was primarily for professionals or elites.[29]

The qualified self wedding reveals the importance of the postdigital within media accounting. The object-oriented ontology is an essential mode through which the wedding exists after the fact. The incorporation of digital traces is recognition of their significance in the ritual itself but their material manifestation is a reclaiming of ownership for the couple. The collective nature in which they were produced mirrors that of the wedding ritual itself but the singularity of the material book provides increased value to the newlywed couple. Here again the postdigital is a move to counter the ephemerality of the digital media environment, but one which quickly becomes entangled in consumer culture.

Implications of Media Accounting

Rethinking Affordances

While "affordance" is a particularly slippery term within contemporary communication and media studies literature, it is nonetheless a common framework for understanding and comparing media technologies. This kind of framework suggests that the technological and design features of media technologies matter in how people can and do use them.[30] While affordances have become a primary way through which we understand and compare digital technologies, it is less often applied to analogue media.

Indeed, analogue media reveal particular affordances of digital media which are often taken for granted. For example, a mobile phone may be about the same size a pocket diary, but it is not squishable or bendable the way a pocket diary is. Therefore, we cannot easily mail a phone in an envelope to someone we love. We would need to put it in a box to mail it, because if it gets squished, it can break. This is in part why we need to buy cases for it. Phones can break if dropped. While you can put a smartphone in your pocket to carry with you, it is difficult to sit down with it in your pocket because it is not bendable. But perhaps the fact that we cannot easily sit with a phone our pockets increases its shareability in some ways.[31] Phones increasingly rest on the table between people, ready at hand to show a picture or to snap a group selfie. Relative affordances across media technologies can reveal constraining characteristics or perceptions thereof which may be liberating in some ways.

For example, Joseph Bayer and his colleagues found that Snapchat affords greater ephemerality than does Facebook.[32] Here the constraint of messages lasting only three to ten seconds enables people to share unflattering or silly pictures with close connections in ways that they would feel inhibited to do on Facebook. The relative comparison of attributes across platforms enables Bayer and his colleagues to suggest that ephemerality is what draws people to Snapchat. While such differences might accurately depict how people experience Snapchat, ephemerality as an attribute might be misleading. As Gaver points out, there are false affordances to technology.[33] Sometimes we perceive something as ephemeral, but it isn't. Screenshots and Snapchat's memories help us to keep Snapchat messages which otherwise "expire." Similarly, photos on a lost mobile phone feel ephemeral only in their loss. Fundamentally, it is important to understand affordances as both relative to others and situated in particular contexts rather than inherent characteristics of a medium or platform.

In additional to the relativity of platform affordances, I want to bring special attention to a very important affordance that is at work throughout the various examples and cases in this book. That is, the *mobility* of media accounting features prominently throughout this work. Not only is the smartphone the primary device through which people engage with various social media, but historically much of media have been mobile (particularly once we moved beyond writing on stone). Harold Innis argues how important the transportability of media was to culture, and thus the transport of codified law and economic structures.[34] While I too want to emphasize the mobility of media throughout this book, I also want to suggest that such mobile media today are also highly individualized. Mimi Ito was the first to articulate the portable, personal, and pedestrian nature of mobile phones.[35] While she used the term pedestrian to refer to an on-the-move characteristic, it certainly also connotes an ordinary and everyday nature. Mobile phones are like other media accounting platforms in the nineteenth century, which were often coupled to a body, a person—in pockets and purses, on belts and in hands. The mobility and embodiment of these technologies shapes the intimacy of media accounting practices.[36] Therefore portable, personal, and pedestrian are affordances that can be used to describe many media accounting platforms across time. One of the challenges of looking at affordances of specific digital media platforms is that they seem unique. We cannot always see the relativity and situatedness of such affordances until we compare across media.

Rethinking Narcissism

The increased visibility of media accounting due to digitalization has led to concerns about increased narcissism more broadly. While media accountings have always circulated within interpersonal networks, social media platforms extend the visibility of our weaker ties' media accounting and beyond. As a result, we have a better sense of others' qualified selves. We also can encounter the media accounting of strangers with ease. Social media platforms expand the visibility of our qualified selves beyond other citizens to include state and corporate actors. Increases in visibility of qualified selves are different from the claims of newness. The qualified self is not a narcissistic self. Part of what it means to be here on earth means to create traces of ourselves for our future selves as well as for those who come after us. Media accounting reveals that we have long created media about ourselves to connect with others.

José van Dijck argues that connectivity is a defining characteristic of social media. Social media, she argues, is not defined by social connectedness, which may be a common value or motivation for many to use such technologies, or at least how the social media companies themselves are articulating their value to users. Instead, van Dijck argues that connectivity is the key characteristic of these socio-technical systems. This connectivity is the result of the database structures and code where connections between data points are computationally made and infinitized. This connectivity is the dominant cultural transformation of social media. We live in a culture of connectivity, she writes, "where perspectives, expressions, experiences and productions are increasingly mediated by social media sites."[37]

While the networked, computational, and algorithmic nature of the digital environment transforms our culture as it presents it back to us, I have shown how all media accounting can do this throughout this book. We should not write about social media platforms as if they are the only sites for social interaction when they exist in a much broader ecosystem of media that facilitate sociality. The practices of media accounting reveal the longstanding ways that media have connected us across time and space. Mediatization reveals, however, the increased role of media institutions and companies into the processes of social connectivity.

Feminization of Culture and Media Accounting

The visibility and prominence of the qualified selves today can be understood through cultural feminization. Lisa Adkins argues the contemporary cultural feminization refers to "a new sovereignty of appearance, image, and style at work, where the performance of stylized presentations of self has emerged as a key resource in certain sectors of the economy, particularly in new service occupations."[38] The qualified self has become not just the work within the domestic sphere but has come to shape expectations within professional spheres as well. Alice Marwick's research on the tech and media elite in Silicon Valley revealed that much of their work was to create and manage professional yet expert mediated representations of self.[39] Cultural feminization would suggest we expect people to have and to manage their qualified selves as part of their professional lives.

More broadly, we see a cultural feminization of media accounting whereby we expect people to share the ordinary aspects of their lives, to perform their multiple identities, to create and engage with remembrances for special events, and to reckon and reconcile their media traces as part of their qualified self. The reflexivity of our media accounting has become part of contemporary expectations for social interaction. If someone doesn't have social media accounts through which we can come to understand them, that is, their qualified self, we wonder, "What's wrong with them? What are they hiding?" Increasingly, there are normative expectations to engage in media accounting to reveal our qualified selves to us as well as the world around us so that we can be read and understood by others.

Media Accounting and Culture

Raymond Williams reminds us that culture can be thought of as divided into the ideal, the documentary, and the social.[40] The ideal culture is that which we aspire to. The documentary culture is the textual, artistic, intellectual, and artifactual products of a society. The social culture is the particular way of life and the everyday practices which represent a society. Media accounting represents all three aspects of culture. Like all media practices, they are simultaneously documentary culture and social culture. But media accounting also reveals the ideal culture as people convey individual and collective aspirations.

The qualified self can also be understood through Williams's cultural triad. The qualified self is the ideal version of ourselves—that which reveals our best intentions and qualities. The qualified self is also the documented version of ourselves. It is not only the textual and visual representations of ourselves but it provides documentary evidence of who we are and what we do. The qualified self is also part of everyday practice. It is part of the ordinary ways that we use media to share with others, to represent our identity work, to create mediated memories, and to understand and consider ourselves in this world. The culture of media accounting gives meaning to who we are and how we live. And it always has.

Notes

Preface

1. Mor Naaman, Jeffrey Boase, and Chih-Hui Lai, "Is It Really About Me? Message Content in Social Awareness Streams," *Proceedings of the ACM Conference on Computer Supported Collaborative Work (CSCW2010)* (New York: Association for Computing Machinery, 2010), 191.

2. Joel Stein, "Millennials: The Me Me Me Generation," *Time*, May 9, 2015, http://time.com/247/millennials-the-me-me-me-generation/.

3. Alex Schrader, "Teens, Social Media and Narcissism," Edudemic, January 30, 2017, http://www.edudemic.com/teens-social-media-narcissism/.

4. Jürgen Habermas, *The Transformation of the Public Sphere: An Inquiry into a Category of Bourgeois Society*, trans. Thomas Burger (Cambridge, MA: MIT Press, 1962), 16–24; Richard Sennett, *The Fall of Public Man* (London: Norton, 1976), 16–27.

Chapter 1: Introduction

1. Margo Culley, ed., *A Day at a Time: The Diary Literature of American Women from 1764 to the Present* (New York: The Feminist Press at CUNY, 1985), 29–35.

2. Drinker's diary is archived in *North American Women's Letters and Diaries: Colonial to 1950*, https://alexanderstreet.com/products/north-american-womens-letters-and-diaries/.

3. Culley, *A Day at a Time*, 3–17.

4. There is a growing interest in media studies, science and technology studies, and information science in understanding the role of historical technologies as potential antecedents to the current social media environment. For example, Esther Milne's book *Letters, Postcards, Email: Technologies of Presence* reveals how postcards are an early form of locative media and Jill Walker Rettberg's *Seeing Ourselves* shows how self-portraits are an early form of selfies. It is this literature to which this book contributes.

5. See William W. Gaver, "Technology Affordances" (paper presented at the Proceedings of the SIGCHI conference on human factors in computing systems: Reaching through technology, New Orleans, LA, April 27–May 2, 1991); and Lucas Graves, "Affordances of Blogging: A Case Study in Culture and Technological Effects," *Journal of Communication Inquiry* 31, no. 4 (2007): 331–346. Many communication scholars actively eschew technological determinism as a means of understanding how technological impacts our lives; the concept of affordances offers a softer determinism. See also Lee Humphreys, "Technological Determinism," in *Encyclopedia of Science and Technology Communication*, ed. Susanna Hornig Priest (Thousand Oaks, CA: Sage Publications, 2010), 869–972, for a discussion of hard versus soft technological determinism.

6. Molly McCarthy, "A Pocketful of Days: Pocket Diaries and Daily Record Keeping among Nineteenth-Century New England Women," *New England Quarterly* 73, no. 2 (2000): 282.

7. McCarthy, "A Pocketful of Days," 295.

8. McCarthy, 295.

9. Jonathan Donner, *After Access: Inclusion, Development, and a More Mobile Internet* (Cambridge, MA: MIT Press, 2015).

10. See Nancy K. Baym, *Personal Connections in the Digital Age* (Malden, MA: Polity Press, 2010); danah boyd, *It's Complicated: The Social Lives of Networked Teens* (New Haven, CT: Yale University Press, 2014); Alice E. Marwick, *Status Update: Celebrity and Attention in the Social Media Age* (New Haven, CT: Yale University Press, 2013); Sherry Turkle, *Alone Together: Why We Expect More from Technology and Less from Each Other* (New York: Basic Books, 2011).

11. Zizi Papacharissi, "The Virtual Sphere 2.0: The Internet, the Public Sphere, and Beyond," in *Handbook of Internet Politics*, eds. Andrew Chadwick and Phil Howard (Abingdon, UK: Routledge, 2008), 230–245.

12. See Clarissa C. David, Jonathan Corpus Ong, and Erika Fille T. Legara, "Tweeting Supertyphoon Haiyan: Evolving Functions of Twitter during and after a Disaster Event," *PLoS One* 11, no. 3 (2016): e0150190; Philip N. Howard and Muzammil M. Hussain, *Democracy's Fourth Wave?: Digital Media and the Arab Spring* (Oxford: Oxford University Press, 2013); Zizi Papacharissi, *Affective Publics: Sentiment, Technology, and Politics* (Oxford: Oxford University Press, 2015); Zeynep Tufekci and Christopher Wilson, "Social Media and the Decision to Participate in Political Protest: Observations from Tahrir Square," *Journal of Communication* 62, no. 2 (2012): 363–379.

13. Tim Highfield, Stephen Harrington, and Axel Bruns, "Twitter as a Technology for Audiencing and Fandom," *Information Communication and Society* 16, no. 3 (2013): 315–339; Mike Thelwall, Kevan Buckley, and Georgios Paltoglou, "Sentiment

in Twitter Events," *Journal of the American Society for Information Science and Technology* 62, no. 2 (2011): 406–418.

14. Twitter, #Numbers, http://blog.twitter.com/2011/03/numbers.html, accessed June 9, 2017.

15. Ben Highmore, *Ordinary Lives: Studies in the Everyday* (London: Routledge, 2010), 6.

16. Roger Silverstone and Leslie Haddon, "Design and the Domestication of Information and Communication Technologies: Technical Change and Everyday Life," in *Communication by Design. The Politics of Information and Communication Technologies*, eds. Robin Mansel and Roger Silverstone (Oxford: Oxford University Press, 1996), 44–74.

17. Carolyn Marvin, *When Old Technologies Were New: Thinking about Electric Communication in the Late Nineteenth Century* (New York: Oxford University Press, 1988), 5.

18. boyd, *It's Complicated.*

19. Eli Pariser, *The Filter Bubble: How the New Personalized Web Is Changing What We Read and How We Think* (London: Penguin Books, 2012); Joseph Turow, *The Daily You: How the New Advertising Industry Is Defining Your Identity and Your Worth* (New Haven, CT: Yale University Press, 2011).

20. Tarleton Gillespie, Pablo J. Boczkowski, and Kirsten A. Foot, "Introduction," in *Media Technologies: Essays on Communication, Materiality, and Society*, eds. Tarleton Gillespie, Pablo J. Boczkowski, and Kirsten J. Foot (Cambridge, MA: MIT Press, 2014), 1–16.

21. Trevor J. Pinch, and Wiebe E. Bijker, "The Social Construction of Facts and Artefacts: Or How the Sociology of Science and the Sociology of Technology Might Benefit Each Other," *Social Studies of Science* 14, no. 3 (1984): 399–441.

22. Joshua Meyrowitz, *No Sense of Place: The Impact of Electronic Media on Social Behavior* (New York: Oxford University Press, 1985), 331.

23. Klaus Bruhn Jensen, "Media," in *International Encyclopedia of Communication*, edited by Wolfgang Donsbach (Blackwell Publishing, 2008), accessed September 12, 2017, http://doi.org/10.1111/b.9781405131995.2008.x

24. Susan J. Douglas, *Listening In: Radio and the American Imagination* (New York: Times Books, 1999).

25. Lynn Spigel, *Make Room for TV: Television and the Family Ideal in Postwar America* (Chicago: University of Chicago Press, 1992).

26. Marvin, *When Old Technologies Were New.*

27. See Jay David Bolter and Richard Grusin, *Remediation: Understanding New Media* (Cambridge, MA: MIT Press, 1999); Katie Day Good, "From Scrapbook to Facebook: A History of Personal Media Assemblage and Archives," *New Media & Society* 15, no. 4 (2013): 557–573; Ellen Garvey, *Writing with Scissors: American Scrapbooks from the Civil War to the Harlem Renaissance* (New York: Oxford University Press, 2012); Lisa Gitelman, *Always Already New: Media, History, and the Data of Culture* (Cambridge, MA: MIT Press, 2006) and *Paper Knowledge: Toward a Media History of Documents* (Durham, NC: Duke University Press, 2014); Jill Walker Rettberg, *Seeing Ourselves through Technology: How We Use Selfies, Blogs and Wearable Devices to See and Shape Ourselves* (Basingstoke, UK: Palgrave Macmillan, 2014).

28. See Bradley S. Greenberg et al., "Comparing Survey and Diary Measures of Internet and Traditional Media Use," *Communication Reports* 18, no. 1 (2005): 1–8; Robert Kubey, Reed Larson, and Mihaly Czikszentmihalyi, "Experience Sampling Method Applications to Communication Research Questions," *Journal of Communication* 4, no. 3 (1996): 99–120.

29. Leah Lievrouw, "New Media, Mediation, and Communication Study," *Information Communication and Society* 12, no. 3 (2009): 303–325.

30. Nick Couldry, *Media, Society, World: Social Media and Digital Media Practice* (London: Polity, 2012), 33–36.

31. Gwenn Schurgin O'Keeffe and Kathleen Clarke-Pearson, "The Impact of Social Media on Children, Adolescents, and Families," *Pediatrics* 127, no. 4 (2011): 802.

32. Simson Garfinkel and David Cox, "Finding and Archiving the Internet Footprint" (paper presented at the First Digital Lives Research Conference: Personal Digital Archives for the 21st Century, London, UK, February 9–11, 2009).

33. Lisa Gitelman, *Paper Knowledge,* 1–3.

34. José van Dijck, "'You Have One Identity': Performing the Self on Facebook and LinkedIn," *Media Culture & Society* 35, no. 2 (2013): 199–215; Liesbet van Zoonen and Georgina Turner, "Exercising Identity: Agency and Narrative in Identity Management," *Kybernetes* 43, no. 6 (2014): 935–946.

35. Tim Highfield, "News via Voldemort: Parody Accounts in Topical Discussions on Twitter," *New Media & Society* 18, no. 9 (2015): 2028–2045.

36. Stuart R. Geiger, "Bots, Bespoke, Code and the Materiality of Software Platforms," *Information Communication and Society* 17, no. 3 (2014): 342–356.

37. Culley, *A Day at a Time.*

38. Janet Golden and Lynn Weiner, "Reading Baby Books: Medicine, Marketing, Money and the Lives of American Infants," *Journal of Social History* 44, no. 3 (2011): 667–687.

39. Ronald D. Lambert, "The Family Historian and Temporal Orientations towards the Ancestral Past," *Time & Society* 5, no. 2 (1996): 115–143.

40. Erving Goffman, *The Presentation of Self in Everyday Life* (Garden City, NY: Doubleday Anchor Books, 1959), 2.

41. David Stark, *The Sense of Dissonance: Accounts of Worth in Economic Life* (Princeton, NJ: Princeton University Press, 2009), 25.

42. Alison Hearn, "Structuring Feeling: Web 2.0, Online Ranking and Rating, and the Digital 'Reputation' Economy," *ephemera: theory & politics in organizations* 10, no. 3–4 (2010): 421–438.

43. Robert A. Fothergill, *Private Chronicles: A Study of English Diaries* (London: Oxford University Press, 1974).

44. Jane H. Hunter, "Inscribing the Self in the Heart of the Family: Diaries and Girlhood in Late-Victorian America," *American Quarterly* 44, no. 1 (1992): 52.

45. Pierre Bourdieu, *Photography: A Middle-Brow Art*, trans. Shaun Whiteside (Stanford, CA: Stanford University Press, 1996).

46. Hunter, "Inscribing the Self."

47. Richard Chalfen, *Snapshot Versions of Life* (Bowling Green, OH: Bowling Green University Press, 1987).

48. Jean Burgess, "Hearing Ordinary Voices: Cultural Studies, Vernacular Creativity and Digital Storytelling," *Continuum* 20, no. 2 (2006): 201–214.

49. Mark Deuze, *Media Work* (London: Polity Press, 2007); Brooke Erin Duffy, *(Not) Getting Paid to Do What You Love* (New Haven, CT: Yale University Press, 2017); David Hesmondhalgh and Sarah Baker, *Creative Labour: Media Work in Three Cultural Industries* (London: Routledge, 2013); Marwick, *Status Update*.

50. Papacharissi, *Affective Publics*.

51. Joseph Turow, *The Aisles Have Eyes: How Retailers Track Your Shopping, Strip Your Privacy, and Define Your Power* (New Haven, CT: Yale University Press, 2017); Turow, *Daily You*.

52. Goffman, *Presentation of Self*.

53. Leonard Reinecke and Sabine Trepte, "Authenticity and Well-Being on Social Network Sites: A Two-Wave Longitudinal Study on the Effects of Online Authenticity and the Positivity Bias in SNS Communication," *Computers in Human Behavior* 30 (2014): 95–102.

54. Peter Sheridan Dodds et al., "Human Language Reveals a Universal Positivity Bias," *Proceedings of the National Academy of Sciences of the United States of America* 112, no. 8 (2015): 2389.

55. Harry F. Wolcott, *Transforming Qualitative Data: Description, Analysis, and Interpretation* (Thousand Oaks, CA: Sage, 1994).

56. George Herbert Mead, *Mind, Self and Society from the Standpoint of a Social Behaviorist* (Chicago: University of Chicago Press, 1934), 135.

57. Mead, *Mind, Self and Society*, 140.

58. Margo Culley, "'I Look at Me': Self as Subject in the Diaries of American Women," *Women's Studies Quarterly* 17, nos. 3–4 (1989): 15–22.

59. Peter Heehs, *Writing the Self: Diaries, Memoirs, and the History of the Self* (New York: Bloomsbury Publishing USA, 2013).

60. Culley, "'I Look at Me.'"

61. Deborah Lupton, *The Quantified Self* (London: Polity, 2016).

62. Gina Neff and Dawn Nafus, *Self-Tracking* (Cambridge, MA: MIT Press, 2016), 28–35.

63. In *Big Data: A Revolution That Will Transform How We Live, Work, and Think* (Boston: Houghton Mifflin Harcourt, 2013), Viktor Mayer-Schönberger and Kenneth Cukier define data as "a description of something that allows it to be recorded, analyzed, and reorganized."

64. Gitelman, *Paper Knowledge*.

65. Mayer-Schönberger and Cukier, *Big Data*.

66. Neff and Nafus, *Self-Tracking*, 15–21.

67. Datafication is a much larger and more complicated process, aesthetic, and mode within society. I focus on the datafication of self-tracking only, but excellent treatments of datafication can be found in Nick Couldry and Andreas Hepp, *The Mediated Construction of Reality* (Cambridge: Polity, 2017); Kate Crawford, Kate Miltner, and Mary L. Gray, "Critiquing Big Data: Politics, Ethics, Epistemology," *International Journal of Communication* 8 (2014): 1663–1672; José van Dijck, "Datafication, Dataism and Dataveillance: Big Data between Scientific Paradigm and Ideology," *Surveillance & Society* 12, no. 2 (2014): 197–208; and the entire special section of *International Journal of Communication* 8 (2014).

68. Roger Silverstone, "The Sociology of Mediation and Communication," in *The Sage Handbook of Sociology*, eds. Craig Calhoun, Chris Rojek, and Bryan Turner (London: Sage, 2005), 202.

69. Silverstone, "Mediation and Communication," 201.

70. Raymond Williams, *Television: Technology and Cultural Form* (New York: Schocken Books, 1975).

71. Good, *From Scrapbook to Facebook.*

72. See Leslie A. Baxter and Barbara M. Montgomery, *Relating: Dialogues and Dialectics* (New York: Guilford Press, 1996), for an interpersonal dialectical framework and Max Horkheimer and Theodor W. Adorno, "The Culture Industry: Enlightenment as Mass Deception," in *Dialectic of Enlightenment,* eds. Max Horkheimer and Theodor W. Adorno (New York: Continuum, 1998), 120–167, for a dialectical analysis of the culture industry.

73. Nancy Fraser, "Rethinking the Public Sphere," in *Habermas and the Public Sphere,* ed. Craig Calhoun (Cambridge, MA: MIT Press, 1992), 109–142.

74. Susan Miller, *Assuming the Position: Cultural Pedagogy and the Politics of Commonplace Writing* (Pittsburgh, PA: University of Pittsburgh Press, 1998).

75. Judith Butler, *Gender Trouble: Feminism and the Subversion of Identity* (New York: Routledge, 1999); Lana F. Rakow and Laura A. Wackwitz, eds., *Feminist Communication Theory: Selections in Context* (Thousand Oaks, CA: Sage, 2004).

76. Jeanne Boydston, *Home and Work: Housework, Wages, and the Ideology of Labor in the Early Republic* (New York: Oxford University Press, 1990); Ann Oakley, *Woman's Work: The Housewife, Past and Present* (New York: Vintage, 1974).

77. Thorstein Veblen, *The Theory of the Leisure Class: An Economic Study of Institutions* (New York: Modern Library, 1934).

78. Arlie Hoschild and Anne Machung, *The Second Shift: Working Parents and the Revolution at Home* (New York: Viking, 1989); Pamela Odih, "Gendered Time in the Age of Deconstruction," *Time & Society* 8, no. 1 (1999): 9–38.

79. Bernie Hogan, "The Presentation of Self in the Age of Social Media: Distinguishing Performances and Exhibitions Online," *Bulletin of Science, Technology & Society* 30, no. 6 (2010): 377–386.

80. José van Dijck, *Mediated Memories in the Digital Age* (Stanford, CA: Stanford University Press, 2007).

81. Jacques Derrida, *Of Grammatology,* corrected ed. (Baltimore: Johns Hopkins University Press, 1997).

Chapter 2: Sharing the Everyday

1. Steve Kelley, "It's a tweet from Amanda …," *Times Picayune,* April 9, 2009, http://editorialcartoonists.com/cartoon/display.cfm/69138/.

2. James W. Carey, *Communication as Culture: Essays on Media and Society* (Boston: Unwin Hyman, 1988), 18–23.

3. Robert A. Fothergill, *Private Chronicles: A Study of English Diaries* (London: Oxford University Press, 1974).

4. Fothergill, *Private Chronicles*, 14.

5. Margo Culley, ed., *A Day at a Time: The Diary Literature of American Women from 1764 to the Present* (New York: The Feminist Press at CUNY, 1985); and Susan Miller, *Assuming the Position: Cultural Pedagogy and the Politics of Commonplace Writing* (Pittsburgh, PA: University of Pittsburgh Press, 1998).

6. Jane H. Hunter, "Inscribing the Self in the Heart of the Family: Diaries and Girlhood in Late-Victorian America," *American Quarterly* 44, no. 1 (1992): 54–56.

7. Robert A. Fothergill, "One Day at a Time: The Diary as Lifewriting," *a/b: Auto/Biography Studies* 10, no. 1 (1995): 90.

8. Culley, *A Day at a Time*.

9. Fothergill, "One Day at a Time," 82.

10. Fothergill, *Private Chronicles*, 10.

11. Fothergill, "One Day at a Time," 90.

12. Fothergill, 87.

13. Lee Humphreys, Thilo von Pape, and Veronika Karnowski, "Evolving Mobile Media: Uses and Conceptualizations of the Mobile Internet," *Journal of Computer-Mediated Communication* 18, no. 4 (2013): 491–507.

14. Catherine M. Bell, *Ritual Theory, Ritual Practice* (New York: Oxford University Press, 2009).

15. Bell, *Ritual Theory, Ritual Practice*, 16.

16. Ien Ang, *Living Room Wars: Rethinking Media Audiences for a Postmodern World* (London: Routledge, 1996); Karin Becker, "Media and the Ritual Process," *Media, Culture & Society* 17, no. 4 (1995): 629–646.

17. Daniel Dayan and Elihu Katz, *Media Events* (Cambridge, MA: Harvard University Press, 1992).

18. Becker, "Media and the Ritual Process," 630, 637.

19. Becker, 640.

20. Carey, *Communication as Culture*, 29.

21. Becker, "Media and the Ritual Process."

22. Lee Humphreys et al., "Historicizing New Media: A Content Analysis of Twitter," *Journal of Communication* 63, no. 3 (2013): 413–431.

23. Heather J. Jackson, *Marginalia: Readers Writing in Books* (New Haven, CT: Yale University Press, 2002).

24. Ken Hillis, *Online a Lot of the Time: Ritual, Fetish, Sign* (Durham, NC: Duke University Press, 2009).

25. Humphreys et al., "Historicizing New Media."

26. Fothergill, *Private Chronicles*; Culley, *A Day at a Time.*

27. Hunter, "Inscribing the Self."

28. Amparo Lasen, "Affective Technologies—Emotions and Mobile Phones," *Receiver* 11 (2004): 2; see also Larissa Hjorth and Sun Sun Lim, "Mobile Intimacy in an Age of Affective Mobile Media," *Feminist Media Studies* 12, no. 4 (2012): 477–484.

29. Jean Burgess and Joshua Green, *YouTube: Online Video and Participatory Culture* (Cambridge: Polity, 2009).

30. Burgess and Green, 53; Nicole Matthews, "Confessions to a New Public: Video Nation Shorts," *Media, Culture & Society* 29, no. 3 (2007): 435–448.

31. Rebecca Steinitz, "Writing Diaries, Reading Diaries: The Mechanics of Memory," *Communication Review* 2, no. 1 (1997): 43–58.

32. Tanya de Grunwald, "Meet the YouTube Big Hitters: The Bright Young Vloggers Who Have More Fans Than 1D," *Daily Mail*, June 14, 2014, http://www.dailymail.co.uk/home/you/article-2656209/The-teen-phenomenon-thats-taking-Youtube.html.

33. de Grunwald, "Meet the YouTube Big Hitters."

34. Maggie Griffith and Zizi Papacharissi, "Looking for You: An Analysis of Video Blogs," *First Monday* 15, nos. 1–4 (2009), http://firstmonday.org/ojs/index.php/fm/article/view/2769/2430/.

35. Sherry Turkle, *Alone Together: Why We Expect More from Technology and Less from Each Other* (New York: Basic Books, 2011).

36. Theresa M. Senft, *Camgirls: Celebrity and Community in the Age of Social Networks* (New York: Peter Lang, 2008); Hillis, *Online a Lot of the Time.*

37. Kate Murphy, "Blogging with Video, Hoping to Go Viral," *New York Times*, February 14, 2013, http://www.nytimes.com/2013/02/14/technology/personaltech/how-to-make-your-video-go-viral.html.

38. Christopher Lasch, *The Culture of Narcissism: American Life in an Age of Diminishing Expectations* (New York: Warner Books, 1979).

39. Zizi Papacharissi, "The Virtual Sphere 2.0: The Internet, the Public Sphere, and Beyond," in *Handbook of Internet Politics*, eds. Andrew Chadwick and Phil Howard (Abingdon, UK: Routledge, 2008), 236–237.

40. Papacharissi, "The Virtual Sphere 2.0," 237.

41. Majd Abdulghani, "Diary of a Saudi Girl: Karate Lover, Science Nerd ... Bride?" Podcast audio, *All Things Considered*, May 31, 2016, http://www.npr.org/sections/goatsandsoda/2016/05/31/479591225/diary-of-a-saudi-girl-karate-lover-science-nerd-bride/.

42. Fothergill, *Private Chronicles*, 8.

43. Jill Lepore, *Book of Ages: The Life and Opinions of Jane Franklin* (New York: Vintage Books, 2014).

44. See Philip N. Howard and Muzammil M. Hussain, *Democracy's Fourth Wave?: Digital Media and the Arab Spring* (Oxford: Oxford University Press, 2013).

Chapter 3: Performing Identity Work

1. Erving Goffman, *The Presentation of Self in Everyday Life* (Garden City, NY: Doubleday Anchor Books, 1959), 3–6.

2. Erving Goffman, *Interaction Ritual: Essays on Face-to-Face Behavior* (Chicago: Aldine Publishers, 1967).

3. James W. Carey, *Communication as Culture: Essays on Media and Society* (Boston: Unwin Hyman, 1988), 18–23.

4. Carey, *Communication as* Culture, 18.

5. Nancy Thumim, *Self-Representation and Digital Culture* (Basingstoke, UK: Palgrave Macmillan, 2012), 7.

6. Theresa M. Senft and Nancy K. Baym, "What Does the Selfie Say? Investigating a Global Phenomenon," *International Journal of Communication* 9 (2015): 1589, http://ijoc.org/index.php/ijoc/article/view/4067/1387/.

7. Paul Frosh, "The Gestural Image: The Selfie, Photography Theory, and Kinesthetic Sociability," *International Journal of Communication* 9 (2015): 1607–1628, http://ijoc.org/index.php/ijoc/article/view/3146/1388/.

8. Senft and Baym, "What Does the Selfie Say?," 1589.

9. Rick Jervis, "Disney Joins Growing Number of Venues Banning Selfie Sticks," *USA Today*, June 26, 2015, http://www.usatoday.com/story/news/2015/06/26/disney-selfie-stick-ban/29333857/.

10. Senft and Baym, "What Does the Selfie Say?," 1590.

11. See Richard Chalfen, *Snapshot Versions of Life* (Bowling Green, OH: Bowling Green University Press, 1987); Ellen Gruber Garvey, *Writing with Scissors: American*

Scrapbooks from the Civil War to the Harlem Renaissance (New York: Oxford University Press, 2012); and Catherine Zuromskis, *Snapshot Photography: The Lives of Images* (Cambridge, MA: MIT Press, 2013).

12. Nancy Martha West, *Kodak and the Lens of Nostalgia* (Charlottesville: University of Virginia, 2000).

13. Reese V. Jenkins, "Technology and the Market: George Eastman and the Origins of Mass Amateur Photography," *Technology and Culture* 16, no. 1 (1975): 16–18.

14. See Roland Barthes, *Camera Lucida: Reflections on Photography* (New York: Hill and Wang, 1981); Walter Benjamin, "The Work of Art in the Age of Mechanical Reproduction: Enlightenment as Mass Deception," in *Media and Cultural Studies: Keyworks*, eds. Meenakshi Gigi Durham and Douglas Kellner (Malden, MA: Wiley-Blackwell, 2001), 48–70; Charles Sanders Peirce, *Peirce on Signs: Writings on Semiotic by Charles Sanders Peirce*, ed. James Hoopes (Chapel Hill: University of North Carolina Press, 1991); Susan Sontag, *On Photography* (New York: Anchor Books Doubleday, 1989).

15. Karin Becker, "The Aesthetics of Amateur Photography," *Adomus* 15 (1993): 19.

16. Jenkins, "Technology and the Market," 16.

17. Sean McCrum, "Snapshot Photography," *Circa* 60 (November–December 1991): 32–36.

18. McCrum, "Snapshot Photography," 36.

19. Pierre Bourdieu, *Photography: A Middle-Brow Art*, trans. Shaun Whiteside (Stanford, CA: Stanford University Press, 1996).

20. Chalfen, *Snapshot Versions of Life*; Julie Hirsch, *Family Photographs: Content, Meaning, and Effect* (New York: Oxford University Press, 1981); and Zuromskis, *Snapshot Photography*.

21. West, *Lens of Nostalgia*.

22. Hirsch, *Family Photographs*.

23. Ronald D. Lambert, "The Family Historian and Temporal Orientations towards the Ancestral Past," *Time & Society* 5, no. 2 (1996): 120.

24. Ann Oakley, *Woman's Work: The Housewife, Past and Present* (New York: Vintage, 1974).

25. Michael Hardt, "Affective Labor," *boundary 2* 26, no. 2 (1999): 89–100.

26. Elizabeth Yakel, "Seeking Information, Seeking Connections, Seeking Meaning: Genealogists and Family Historians," *Information Research: An International Electronic Journal* 10, no. 1 (2004), http://www.informationr.net/ir/10-1/paper205.html.

27. Becker, "Aesthetics of Amateur Photography," 5.

28. Graham King, *Say "Cheese"! Looking at Snapshots in a New Way* (New York: Dodd, Mead and Company, 1984).

29. Zuromskis, *Snapshot Photography*, 39.

30. Zuromskis, 31.

31. Eden Litt, "Knock, Knock. Who's There? The Imagined Audience," *Journal of Broadcasting & Electronic Media* 56, no. 3 (2012): 331.

32. Nancy K. Baym, *Personal Connections in the Digital Age* (Malden, MA: Polity Press, 2010), 154–155.

33. West, *Lens of Nostalgia*; and Becker, "Aesthetics of Amateur Photography."

34. Bourdieu, *Photography*.

35. West, *Lens of Nostalgia*, plate 8.

36. West, 13.

37. Ellen Gruber Garvey, *The Adman in the Parlor: Magazines and the Gendering of Consumer Culture, 1880s to 1910s* (New York: Oxford University Press, 1996).

38. West, *Lens of Nostalgia*, 1.

39. Katherine Ott, Susan Tucker, and Patricia P. Buckler, "An Introduction to the History of Scrapbooks," in *The Scrapbook in American Life*, eds. Susan Tucker, Katherine Ott, and Patricia P. Buckler (Philadelphia, PA: Temple University Press, 2006), 3.

40. Zuromskis, *Snapshot Photography*, 9.

41. Garvey, *Writing with Scissors*, 20.

42. Garvey, 4.

43. Garvey, 52.

44. Ott, Tucker, and Buckler, "History of Scrapbooks," 3.

45. Garvey, *Writing with Scissors*, 4.

46. Susan Miller, *Assuming the Position: Cultural Pedagogy and the Politics of Commonplace Writing* (Pittsburgh, PA: University of Pittsburgh Press, 1998), 5–6.

47. Garvey, *Writing with Scissors*, 27–28.

48. Ellen Walkley, "OHQ Research Files: Scraps of History: Researching Scrapbooks at the Oregon Historical Society," *Oregon Historical Quarterly* 102, no. 4 (2001): 515.

49. Walkley, "Scraps of History," 515.

50. Humphreys et al., "Historicizing New Media."

51. Leah Scolere and Lee Humphreys, "Pinning Design: The Curatorial Labor of Creative Professionals," *Social Media + Society* 2, no. 1 (2016). http://doi.org/10.1177/2056305116633481.

52. Eric Gilbert et al., "'I Need to Try This!': A Statistical Overview of Pinterest" (paper presented at the Association for Computing Machinery SIGCHI Conference on Human Factors in Computing Systems, Paris, France, April 27–May 2, 2013).

53. Janet Golden and Lynn Weiner, "Reading Baby Books: Medicine, Marketing, Money and the Lives of American Infants," *Journal of Social History* 44, no. 3 (2011): 667.

54. Golden and Weiner, "Reading Baby Books," 672.

55. William Crain, *Theories of Development: Concepts and Applications*, 4th ed. (Upper Saddle River, NJ: Prentice Hall, 2000), 21.

56. Golden and Weiner, "Reading Baby Books," 667, 673.

57. Kevin McGee and Jörgen Skågeby, "Gifting Technologies," *First Monday* 9, no. 12 (2004), http://firstmonday.org/ojs/index.php/fm/article/view/1192/1112/.

58. Kazys Varnelis, ed., *Networked Publics* (Cambridge, MA: MIT Press, 2008), 5.

59. Silvia Federici, *Wages against Housework* (Bristol, UK: Power of Women Collective and Falling Wall Press, 1975).

60. Garvey, *Writing with Scissors*.

61. Liesbet van Zoonen and Georgina Turner, "Exercising Identity: Agency and Narrative in Identity Management," *Kybernetes* 43, no. 6 (2014): 935–946.

62. Jefferson Pooley, "The Consuming Self: From Flappers to Facebook," in *Blowing up the Brand: Critical Perspectives on Promotional Culture*, eds. Melissa Aronczyk and Devon Powers (New York: Peter Lang, 2010), 79–80.

63. Jeanne Boydston, *Home and Work: Housework, Wages, and the Ideology of Labor in the Early Republic* (New York: Oxford University Press, 1990).

64. Judith Donath and danah boyd, "Public Displays of Connection," *BT Technology Journal* 22, no. 4 (2004): 71–82; Alice E. Marwick and danah boyd, "I Tweet Honestly, I Tweet Passionately: Twitter Users, Context Collapse, and the Imagined Audience," *New Media & Society* 13, no. 1 (2011): 114–133.

65. Van Zoonen and Turner, "Exercising Identity," 937.

66. Jessica Vitak, "'Why Won't You Be My Facebook Friend?': Strategies for Managing Context Collapse in the Workplace" (proceedings of the 2012 iConference, Toronto, Ontario, Canada: ACM, 2012), 556.

67. Marwick and boyd, "I Tweet Honestly," 125.

68. Joseph Turow, *The Daily You: How the New Advertising Industry Is Defining Your Identity and Your Worth* (New Haven, CT: Yale University Press, 2011).

69. Goffman, *Interaction Ritual.*

70. Tiziana Terranova, "Free Labor: Producing Culture for the Digital Economy," *Social Text* 18, no. 2 (2000): 33–34.

Chapter 4: Remembrancing

1. Annette Kuhn, "Memory Texts and Memory Work: Performances of Memory in and with Visual Media," *Memory Studies* 3, no. 4 (2010): 303.

2. José van Dijck, *Mediated Memories in the Digital Age* (Stanford, CA: Stanford University Press, 2007), 21.

3. Christine Lohmeier and Christian Pentzold, "Making Mediated Memory Work: Cuban-Americans, Miami Media and the Doings of Diaspora Memories," *Media, Culture & Society* 36, no. 6 (2014): 778.

4. Lohmeier and Pentzold, "Making Mediated Memory Work," 780.

5. Martin A. Conway and Susan E. Gathercole, "Writing and Long-Term Memory: Evidence for a 'Translation' Hypothesis," *Quarterly Journal of Experimental Psychology Section A* 42, no. 3 (1990): 521.

6. Esther Milne, *Letters, Postcards, Email: Technologies of Presence* (New York: Routledge, 2010), 202.

7. "What Is Snapchat?" YouTube video, 3:56, posted by "Snapchat," June 16, 2015, https://youtu.be/ykGXIQAHLnA.

8. Valerie Payne, "Grieving Mom Refuses to Remove Photos of Stillborn Son from Facebook," *HLN*, October 5, 2015, http://www.hlntv.com/articles/2015/10/05/mom-posts-photos-of-stillborn-son-on-facebook-sparks-controversy/; Britney Glaser, "Mother Will Continue Sharing Photos of Her Stillborn Son, Despite Backlash," KPLC, October 1, 2015, http://www.kplctv.com/story/30164909/mother-will-continue-sharing-photos-of-her-stillborn-son-despite-backlash.

9. Valerie Payne, "Grieving Mom."

10. Stanley B. Burns, *Sleeping Beauty III: Memorial Photography: The Children* (New York: Burns Archive Press, 2011).

11. Burns, *Sleeping Beauty III*; Laurel Hilliker, "Letting Go While Holding On: Postmortem Photography as an Aid in the Grieving Process," *Illness, Crisis & Loss* 14, no. 3 (2006): 245–269.

12. Hilliker, "Letting Go," 254.

13. Denise McGuinness, Barbara Coughlan, and Sheila Power, "Empty Arms: Supporting Bereaved Mothers during the Immediate Postnatal Period," *British Journal of Midwifery* 22, no. 4 (2014): 251.

14. Cybele Blood and Joanne Cacciatore, "Best Practice in Bereavement Photography after Perinatal Death: Qualitative Analysis with 104 Parents," *BMC Psychology* 2 (2014): Article 15.

15. Cybele Blood and Joanne Cacciatore, "Parental Grief and Memento Mori Photography: Narrative, Meaning, Culture, and Context," *Death Studies* 38, no. 4 (2014): 224–233; Katherine J. Gold, Vanessa K. Dalton, and Thomas L. Schwenk, "Hospital Care for Parents after Perinatal Death," *Obstetrics and Gynecology* 109, no. 5 (2007): 1156–1166.

16. Carole Novielli, "Remembering Grayson: Anencephalic Baby Facebook Banned Whose Life Impacted So Many," Life News, March 3, 2014, http://www.lifenews.com/2014/03/03/remembering-grayson-anencephalic-baby-facebook-banned-whose-life-impacted-so-many/. (Please note that this article contains difficult imagery.)

17. Hilliker, "Letting Go," 254.

18. Gold, Dalton, and Schwenk, "Hospital Care for Parents."

19. Hilliker, "Letting Go," 246.

20. Barbie Zelizer, *Remembering to Forget: Holocaust Memory through the Camera's Eye* (Chicago: University of Chicago Press, 1998), 8–10.

21. Kenza Moller, "Why You Shouldn't Post Photos of 9/11 on Social Media." Romper, September 11, 2016, https://www.romper.com/p/why-you-shouldnt-post-photos-from-911-on-social-media-18144/.

22. Eric Meyer, "Inadvertent Algorithmic Cruelty," personal website, December 24, 2014, http://meyerweb.com/eric/thoughts/2014/12/24/inadvertent-algorithmic-cruelty/.

23. Tarleton Gillespie, "The Relevance of Algorithms," in *Media Technologies: Essays on Communication, Materiality, and Society*, eds. Tarleton Gillespie, Pablo Boczkowski, and Kirsten A. Foot (Cambridge, MA: MIT Press, 2014), 175.

Chapter 5: Reckoning

1. Joshua Meyrowitz, "Watching Us Being Watched: State, Corporate, and Citizen Surveillance" (paper presented at the symposium *The End of Television? Its Impact on the World (So Far)* Annenberg School for Communication, University of Pennsylvania, Philadelphia, February 17–18, 2007), 1–3.

2. Jill Walker Rettberg, *Seeing Ourselves through Technology: How We Use Selfies, Blogs and Wearable Devices to See and Shape Ourselves* (Basingstoke, UK: Palgrave Macmillan, 2014).

3. Michel Foucault, "Technologies of the Self," in *Technologies of the Self: A Seminar with Michel Foucault*, eds. Luther H. Martin, Huck Gutman, and Patrick H. Hutton (Amherst, MA: University of Massachusetts Press, 1988), 16–49.

4. Robert A. Fothergill, *Private Chronicles: A Study of English Diaries* (London: Oxford University Press, 1974).

5. Jane H. Hunter, "Inscribing the Self in the Heart of the Family: Diaries and Girlhood in Late-Victorian America," *American Quarterly* 44, no. 1 (1992): 51–81.

6. Herman M. Serota, "Home Movies of Early Childhood: Correlative Developmental Data in the Psychoanalysis of Adults," *Science* 143, no. 3611 (1964): 1195.

7. See J. L. Adrien et al., "Autism and Family Home Movies: Preliminary Findings," *Journal of Autism and Developmental Disorders* 21, no. 1 (1991): 43–49.

8. See Werner F. Helsen and Janet L. Starkes, "A Multidimensional Approach to Skilled Perception and Performance in Sport," *Applied Cognitive Psychology* 13, no. 1 (1999): 1–27.

9. Paul Messaris, *Visual "Literacy": Image, Mind, and Reality* (Boulder, CO: Westview Press, 1994).

10. Messaris, *Visual Literacy*.

11. H. P. Grice, "Logic in Conversation," in *Syntax and Semantics, Vol 3: Speech Acts*, eds. Peter Cole and Jerry L. Morgan (New York: Academic Press, 1972), 46.

12. Fothergill, *Private Chronicles*.

13. Alessandro Acquisti and Christina M. Fong, "An Experiment in Hiring Discrimination via Online Social Networks," July 17, 2015, http://doi.org/10.2139/ssrn.2031979.

14. Russ Buettner, "A Brooklyn Protester Pleads Guilty after His Twitter Posts Sink His Case," *New York Times*, December 12, 2012, http://nytimes.com/2012/12/13/nyregion/malcolm-harris-pleads-guilty-over-2011-march.html.

15. Julie Hirsch, *Family Photographs: Content, Meaning, and Effect* (New York: Oxford University Press, 1981), 12–13.

16. Matt Raymond, "How Tweet It Is!: Library Acquires Entire Twitter Archive," Library of Congress, April 14, 2010, http://blogs.loc.gov/loc/2010/04/how-tweet-it-is-library-acquires-entire-twitter-archive/.

17. Hayley Tsukayama, "Fueled by Stoke, GoPro Helps Turn Rookie Selfie Snappers into Film Directors," *Washington Post*, October 10, 2014, http://wapo.st/1wb0qSu/.

18. Gabe Johnson and Nick Wingfield, "GoPro Goes Amateur," *New York Times*, January 31, 2014, https://www.nytimes.com/2014/01/31/technology/gopro-works-on-its-brand.html.

19. GoPro, "2015 Annual Financial Report,"2016, https://s21.q4cdn.com/291350743/files/doc_financials/2015/GoPro_-_2015_Annual_Report.pdf.

20. "GoPro Awards: 14-Year-Old Shreds Chopin Piano Solo in 4K," YouTube video, 1:10, posted by "GoPro," April 11, 2016, https://www.youtube.com/watch?v=AyuAHa3JVfc.

21. Meyrowitz, "Watching Us Being Watched," 1–3.

22. See also Lee Humphreys, "Who's Watching Whom? A Study of Interactive Technology and Surveillance," *Journal of Communication* 61, no. 4 (2011): 557–558, http://doi.org/10.1111/j.1460-2466.2011.01570.x.

23. Theresa M. Senft and Nancy K. Baym, "What Does the Selfie Say? Investigating a Global Phenomenon," *International Journal of Communication* 9 (2015): 1588–1606; Rettberg, *Seeing Ourselves through Technology*, 36.

24. Kath Albury, "Selfies, Sexts and Sneaky Hats: Young People's Understandings of Gendered Practices of Self-Representation," *International Journal of Communication* 9 (2015): 1740–1741.

25. Nick Couldry, *Media, Society, World: Social Media and Digital Media Practice* (London: Polity, 2012), 49–51.

26. Ingrid Richardson and Rowan Wilken, "Parerga of the Third Screen: Mobile Media, Place, Presence," in *Mobile Technology and Place*, eds. Rowan Wilken and Gerard Goggin (New York: Routledge, 2012), 194.

27. David Nemer and Guo Freeman, "Empowering the Marginalized: Rethinking Selfies in the Slums of Brazil," *International Journal of Communication* 9 (2015): 1839.

28. James Meese et al., "Selfies at Funerals: Mourning and Presencing on Social Media Platforms," *International Journal of Communication* 9 (2015): 1819.

29. Rettberg, *Seeing Ourselves through Technology*, 35–40.

30. In *Seeing Ourselves through Technology*, Rettberg makes a similar distinction between cumulative and whole. She argues digital self-presentations are never whole, only cumulative.

31. Derrida, *On Grammatology*, 46–48.

32. Geri Gay, *Context-Aware Mobile Computing: Affordances of Space, Social Awareness, and Social Influence* (San Rafael, CA: Morgan and Claypool, 2009).

33. Foucault, "Technologies of the Self"; Rettberg, *Seeing Ourselves through Technology*.

34. Gina Neff and Dawn Nafus, *Self-Tracking* (Cambridge, MA: MIT Press, 2016); Deborah Lupton, *The Quantified Self* (London: Polity, 2016).

35. Edward E. Sampson, "The Deconstruction of the Self," in *Texts of Identity*, eds. John Shotter and Kenneth J. Gergen (London: Sage, 1989), 13.

36. Derrida, *On Grammatology*.

37. Neff and Nafus, *Self-Tracking*, 89–92.

38. Michael Mayer, dir., *Mortified Nation* [Documentary] (Los Angeles: Wiser Post, 2013).

39. Linda Flanagan, "Reflecting on Adolescence: How Stories Can Inspire Teen Empathy," KQED News, June 15, 2016, https://kqed.org/mindshift/2016/06/15/mortified-how-sharing-childhood-journals-can-help-teens-feel-more-understood/.

40. Limor Shifman, *Memes in Digital Culture* (Cambridge, MA: MIT Press, 2014).

41. Elise Moreau, "What Throwback Thursday Actually Is and Why It's So Popular," *Life Wire*, August 14, 2017, https://www.lifewire.com/throwback-thursday-meaning-and-why-its-so-popular-3485860.

42. Steven Petrow, "Navigating the Rules of 'Throwback Thursday,'" *USA Today*, August 20, 2015, https://www.usatoday.com/story/tech/personal/2015/08/20/navigating-rules-throwback-thursday/32041277/.

43. Brian Koerber, "An Unofficial Guide for #ThrowbackThursday Etiquette," *Mashable*, April 17, 2014, http://mashable.com/2014/04/17/throwback-thursday-etiquette/.

44. Robert A. Fothergill, "One Day at a Time: The Diary as Lifewriting," *a/b: Auto/Biography Studies* 10, no. 1 (1995): 87.

Chapter 6: Conclusion

1. See Karine Nahon and Jeff Hemsley, *Going Viral* (Cambridge: Polity, 2013).

2. Judy Wajcman, *Pressed for Time: The Acceleration of Life in Digital Capitalism* (Chicago: University of Chicago Press, 2015).

3. Michel Foucault, "Technologies of the Self," in *Technologies of the Self: A Seminar with Michel Foucault*, eds. Luther H. Martin, Huck Gutman, and Patrick H. Hutton (Amherst, MA: University of Massachusetts Press, 1988), 18–19; Deborah Lupton, *The Quantified Self* (London: Polity, 2016); Jill Walker Rettberg, *Seeing Ourselves through Technology: How We Use Selfies, Blogs and Wearable Devices to See and Shape Ourselves* (Basingstoke, UK: Palgrave Macmillan, 2014), 83–84.

4. Nick Couldry and Andreas Hepp, *The Mediated Construction of Reality* (Cambridge: Polity, 2017), 35.

5. Of course, in *Orality and Literacy* (London: Routledge, 2012), 115–116, Walter Ong reminds us that mechanization is coupled with other necessary social changes, such as the Protestant reformation and increasing literacy rates more broadly from 1440 to 1840 that enable media accounting.

6. Molly A. McCarthy, *The Accidental Diarist: A History of the Daily Planner in America* (Chicago: University of Chicago Press, 2013), 7–9.

7. In addition to the journals other technologies brought about media accounting. Ong (*Orality and Literacy*, 80–82) argues that writing itself can be understood as a technology. David M. Henkin, *The Postal Age: The Emergence of Modern Communications in Nineteenth-Century America* (Chicago: University of Chicago Press, 2007) also describes how the nineteenth-century postal system enabled the networked circulation of media across geographical space. Therefore, we can see a complex technological system of writing, paper production, and book making, as well as the mail system, which contributed to and shaped the practices of media accounting.

8. Couldry and Hepp, *Mediated Construction of Reality*, 38–40.

9. Richard Chalfen, *Snapshot Versions of Life* (Bowling Green, OH: Bowling Green University Press, 1987), 8–9. See also Molly A. McCarthy's *The Accidental Diarist* as an excellent example of scholarship that has extensively engaged with diaries and daily planners as media.

10. Axel Bruns, *Blogs, Wikipedia, Second Life, and Beyond: From Production to Produsage* (New York: Peter Lang, 2008), 2–6.

11. danah boyd, "Social Network Sites as Networked Publics: Affordances, Dynamics, and Implications," in *Networked Publics: Identity, Community, and Culture on Social Network Sites*, ed. Zizi Papacharissi (New York: Routledge, 2011), 39–58.

12. The use of webcams to broadcast people's everyday lives are part of self-documentation, but model a distinctive broadcast understanding of mediated everyday life (Theresa Senft, *Camgirls: Celebrity and Community in the Age of Social Networks* [New York: Peter Lang, 2008]) or fetishization of everyday life (Ken Hillis, *Online a Lot of the Time: Ritual, Fetish, Sign* [Durham, NC: Duke University Press, 2009]). The focus of this scholarship was webcam uses for peer broadcasting, not the creation of media traces.

13. This of course becomes particularly funny when people would add details to their profile which occurred in the past but Facebook treated as an announcement because it was new to Facebook. I still remember when my friend Jane included in her profile that she was married to Corey. They had been married in 2002, but

joined Facebook in 2007. When she added the detail to her profile it was as if it was a new relationship. Several people congratulated her.

14. Amanda Keegan, "4 Ways to Save Money on Your Cell Phone Bill," ABC News, May 26, 2015, http://abcnews.go.com/Business/ways-save-money-cell-phone-bill/story?id=31297718. The American mobile telecommunications system typically subsidizes the cost of the phone itself in return for two-year mobile contract commitments.

15. In addition, Facebook structurally refers to itself as "products," which include the profile, newsfeed, messenger, video, photos, search, pages, Instagram, and so on: https://newsroom.fb.com/products, accessed June 28, 2017. On a personal note, I've known several colleagues and students who have gone to work at Facebook and for some reason, I have always been struck by the "product" language with which they describe their work at Facebook. This clearly reflects a late capitalistic logic of production (David Harvey, *The Condition of Postmodernity* [Malden, MA: Blackwell Publishers, 1990], 147–51).

16. Jonathan Zittrain, *The Future of the Internet: And How to Stop It* (New Haven, CT: Yale University Press, 2008), chapter 4. Indeed, Zuckerberg's initiative to bring free "lightweight" mobile Internet access to the world's unconnected populations, entitled Facebook Zero, received strong pushback because it was perceived as a walled garden, as described by Michael Best, "The Internet That Facebook Built" *Communications of the ACM* 57, no. 12 (2014): 21–23

17. Wendy Hui Kyong Chun, *Updating to Remain the Same: Habitual New Media* (Cambridge, MA: MIT Press, 2016).

18. Claude Fischer, *America Calling: A Social History of the Telephone to 1940*. (Berkeley: University of California Press, 1992), 36.

19. Reese V. Jenkins, "Technology and the Market: George Eastman and the Origins of Mass Amateur Photography," *Technology and Culture* 16, no. 1 (1975): 16.

20. Viktor Mayer-Schönberger and Kenneth Cukier, *Big Data: A Revolution That Will Transform How We Live, Work, and Think* (Boston: Houghton Mifflin Harcourt, 2013), 53–61; and Mark Poster, *The Mode of Information: Poststructuralism and Social Construct* (Chicago: University of Chicago Press, 1990), 6–7.

21. Christian Fuchs, "Digital Prosumption Labour on Social Media in the Context of the Capitalist Regime of Time," *Time & Society* 23, no. 1 (2014): 111–113.

22. Couldry and Hepp, *Mediated Construction of Reality*, 38, 122–125.

23. Sy Taffel, "Perspectives on the Postdigital: Beyond Rhetorics of Progress and Novelty," *Convergence* 2, no. 3 (2016): 325.

24. Jonathan Sterne, "Analog" in *Digital Keywords: A Vocabulary of Information Society and Culture*, ed. Benjamin Peters (Princeton, NJ: Princeton University Press, 2016), 31–35.

25. Florian Cramer, "What Is 'Post-Digital'?" *A Peer Reviewed Journal About Post-Digital Research* 13, no. 1 (2014), http://aprja.net/what-is-post-digital/.

26. Prynt, "Introducing the all new Prynt Pocket," https://www.youtube.com/watch?v=tUi7gFgEn8Q, accessed July 6, 2017.

27. Caroline Basset, "Not Now? Feminism, Technology, Postdigital," in *Postdigital Aesthetics: Art, Computation and Design*, eds. David M. Berry and Michael Dieter (London: Palgrave Macmillan, 2015), 138.

28. https://www.mysocialbook.com, accessed 6 July 2017. I'm not sure why they capitalize Book, Life, or Forever.

29. Ellen Gruber Garvey, *Writing with Scissors: American Scrapbooks from the Civil War to the Harlem Renaissance* (New York: Oxford University Press, 2012), 247.

30. See William W. Gaver, "Technology Affordances" (paper presented at the Association for Computing Machinery SIGCHI Conference on Human Factors in Computing Systems: Reaching through Technology, New Orleans, LA, April 27–May 2, 1991), 80–81; Lucas Graves, "The Affordances of Blogging: A Case Study in Culture and Technological Effects," *Journal of Communication Inquiry* 31, no. 4 (2007): 335–338; Ashlee Humphreys, *Social Media: Enduring Principles* (New York: Oxford University Press, 2015), 24–29.

31. Andrew Richard Schrock, "Communicative Affordances of Mobile Media: Portability, Availability, Locatability, and Multimediality," *International Journal of Communication* 9 (2015): 1235–1238, http://ijoc.org/index.php/ijoc/article/view/3288/1363/. Schrock identifies portability, locatability, availability, and multimediality as the affordances of mobile media, but locatability and availability are about being found or being the recipient of calls/texts, whereas I want to stress the active production as well as reception of media accounting.

32. Joseph B. Bayer et al., "Sharing the Small Moments: Ephemeral Social Interaction on Snapchat," *Information, Communication & Society* 19, no. 7 (2016): 958–959.

33. Gaver, *Technology Affordances*, 80.

34. Harold A. Innis, "The Bias of Communication," *Canadian Journal of Economics and Political Science* 15, no. 4 (1949): 467–469.

35. Mizuko Ito, "Introduction: Personal, Portable, Pedestrian," in *Personal, Portable, Pedestrian: Mobile Phones in Japanese Life*, eds. Mizuko Ito, Daisuke Okabe, and Misa Matsuda (Cambridge, MA: MIT Press, 2005), 1–4.

36. Jason Farman, *Mobile Interface Theory: Embodied Space and Locative Media* (New York: Routledge, 2012), 19–31.

37. José van Dijck, "Flickr and the Culture of Connectivity: Sharing Views, Experiences, Memories," *Memory Studies* 4, no. 4 (2010): 402.

38. Lisa Adkins, "Cultural Feminization: "Money, Sex and Power" for Women." *Signs* 26, no. 3 (2001): 674.

39. Alice E. Marwick, *Status Update: Celebrity and Attention in the Social Media Age* (New Haven, CT: Yale University Press, 2013).

40. Raymond Williams, *Culture and Society* (New York: Harper and Row, 1966), 295.

Bibliography

Abdulghani, Majd. "Diary of a Saudi Girl: Karate Lover, Science Nerd ... Bride?" Podcast audio. *All Things Considered,* May 31, 2016. http://www.npr.org/sections/goatsandsoda/2016/05/31/479591225/diary-of-a-saudi-girl-karate-lover-science-nerd-bride/.

Acquisti, Alessandro, and Christina M. Fong. "An Experiment in Hiring Discrimination via Online Social Networks." July 17, 2015. doi:10.2139/ssrn.2031979.

Adkins, Lisa. "Cultural Feminization: 'Money, Sex and Power' for Women." *Signs* 26 (3) (2001): 669–695.

Adrien, J. L., M. Faure, A. Perrot, L. Hameury, B. Garreau, C. Barthelemy, and D. Sauvage. "Autism and Family Home Movies: Preliminary Findings." *Journal of Autism and Developmental Disorders* 21 (1) (1991): 43–49.

Albury, Kath. "Selfies, Sexts and Sneaky Hats: Young People's Understandings of Gendered Practices of Self-Representation." *International Journal of Communication* 9 (2015): 1734–1745.

Ang, Ien. *Living Room Wars: Rethinking Media Audiences for a Postmodern World.* London: Routledge, 1996.

Barthes, Roland. *Camera Lucida: Reflections on Photography.* New York: Hill and Wang, 1981.

Basset, Caroline. "Not Now? Feminism, Technology, Postdigital." In *Postdigital Aesthetics: Art, Computation and Design,* edited by David M. Berry and Michael Dieter. 136–150. London: Palgrave Macmillan, 2015.

Baxter, Leslie A., and Barbara M. Montgomery. *Relating: Dialogues and Dialectics.* New York: Guilford Press, 1996.

Bayer, Joseph B., Nicole B. Ellison, Sarita Y. Schoenebeck, and Emily B. Falk. "Sharing the Small Moments: Ephemeral Social Interaction on Snapchat." *Information, Communication & Society* 19 (7) (2016): 956–977.

Baym, Nancy K. *Personal Connections in the Digital Age*. Malden, MA: Polity Press, 2010.

Becker, Karin. "The Aesthetics of Amateur Photography." *Adomus* 15 (1993): 19–25.

Becker, Karin. "Media and the Ritual Process." *Media, Culture & Society* 17 (4) (1995): 629–646.

Bell, Catherine M. *Ritual Theory, Ritual Practice*. New York: Oxford University Press, 2009.

Benjamin, Walter. "The Work of Art in the Age of Mechanical Reproduction: Enlightenment as Mass Deception." In *Media and Cultural Studies: Keyworks*. Edited by Meenakshi Gigi Durham and Douglas Kellner, 48–70. Malden, MA: Wiley-Blackwell, 2001.

Best, Michael L. "The Internet That Facebook Built." *Communications of the ACM* 57 (12) (2014): 21–23.

Blood, Cybele, and Joanne Cacciatore. "Best Practice in Bereavement Photography after Perinatal Death: Qualitative Analysis with 104 Parents." *BMC Psychology* 2 (2014): Article 15. http://www.biomedcentral.com/2050-7283/2/15/.

Blood, Cybele, and Joanne Cacciatore. "Parental Grief and Memento Mori Photography: Narrative, Meaning, Culture, and Context." *Death Studies* 38 (4) (2014): 224–233.

Bolter, Jay David, and Richard Grusin. *Remediation: Understanding New Media*. Cambridge, MA: MIT Press, 1999.

Bourdieu, Pierre. *Photography: A Middle-Brow Art*. Translated by Shaun Whiteside. Stanford, CA: Stanford University Press, 1996.

boyd, danah. *It's Complicated: The Social Lives of Networked Teens*. New Haven, CT: Yale University Press, 2014.

boyd, danah. "Social Network Sites as Networked Publics: Affordances, Dynamics, and Implications." In *Networked Publics: Identity, Community, and Culture on Social Network Sites*, edited by Zizi Papacharissi, 39–58. New York: Routledge, 2011.

Boydston, Jeanne. *Home and Work: Housework, Wages, and the Ideology of Labor in the Early Republic*. New York: Oxford University Press, 1990.

Bruns, Axel. *Blogs, Wikipedia, Second Life, and Beyond: From Production to Produsage*. New York: Peter Lang, 2008.

Buettner, Russ. "A Brooklyn Protester Pleads Guilty after His Twitter Posts Sink His Case." *New York Times*, December 12, 2012. http://nytimes.com/2012/12/13/nyregion/malcolm-harris-pleads-guilty-over-2011-march.html.

Burgess, Jean. "Hearing Ordinary Voices: Cultural Studies, Vernacular Creativity and Digital Storytelling." *Continuum* 20 (2) (2006): 201–214.

Burgess, Jean, and Joshua Green. *YouTube: Online Video and Participatory Culture*. Cambridge: Polity, 2009.

Burns, Stanley B. *Sleeping Beauty III: Memorial Photography: The Children*. New York: Burns Archive Press, 2011.

Butler, Judith. *Gender Trouble: Feminism and the Subversion of Identity*. New York: Routledge, 1999.

Carey, James W. *Communication as Culture: Essays on Media and Society*. Boston: Unwin Hyman, 1988.

Chalfen, Richard. *Snapshot Versions of Life*. Bowling Green, OH: Bowling Green University Press, 1987.

Chun, Wendy Hui Kyong. *Updating to Remain the Same: Habitual New Media*. Cambridge, MA: MIT Press, 2016.

Conway, Martin A., and Susan E. Gathercole. "Writing and Long-Term Memory: Evidence for a 'Translation' Hypothesis." *Quarterly Journal of Experimental Psychology Section A* 42 (3) (1990): 513–527.

Couldry, Nick. *Media, Society, World: Social Media and Digital Media Practice*. London: Polity, 2012.

Couldry, Nick, and Andreas Hepp. *The Mediated Construction of Reality*. Cambridge: Polity, 2017.

Crain, William. *Theories of Development: Concepts and Applications*. 4th ed. Upper Saddle River, NJ: Prentice Hall, 2000.

Cramer, Florian. "What Is 'Post-Digital'?" *A Peer Reviewed Journal About Post-Digital Research* 13 (1) (2014). http://aprja.net/what-is-post-digital/.

Crawford, Kate, Kate Miltner, and Mary L. Gray. "Critiquing Big Data: Politics, Ethics, Epistemology." *International Journal of Communication* 8 (2014): 1663–1672.

Culley, Margo, ed. *A Day at a Time: The Diary Literature of American Women from 1764 to the Present*. New York: The Feminist Press at CUNY, 1985.

Culley, Margo. "'I Look at Me': Self as Subject in the Diaries of American Women." *Women's Studies Quarterly* 17 (3–4) (1989): 15–22.

David, Clarissa C., Jonathan Corpus Ong, and Erika Fille T. Legara. "Tweeting Supertyphoon Haiyan: Evolving Functions of Twitter During and after a Disaster Event." *PLoS One* 11 (3) (2016): e0150190. doi:10.1371/journal.pone.0150190.

Dayan, Daniel, and Elihu Katz. *Media Events*. Cambridge, MA: Harvard University Press, 1992.

de Grunwald, Tanya. "Meet the YouTube Big Hitters: The Bright Young Vloggers Who Have More Fans Than 1D." *Daily Mail*, June 14, 2014. http://www.dailymail.co.uk/home/you/article-2656209/The-teen-phenomenon-thats-taking-Youtube.html.

Derrida, Jacques. *Of Grammatology*. Corrected ed. Baltimore: Johns Hopkins University Press, 1997.

Deuze, Mark. *Media Work*. London: Polity Press, 2007.

Dodds, Peter Sheridan, Eric M. Clark, Suma Desu, Morgan R. Frank, Andrew J. Reagan, Jake Ryland Williams, Lewis Mitchell, et al. "Human Language Reveals a Universal Positivity Bias." *Proceedings of the National Academy of Sciences of the United States of America* 112 (8) (2015): 2389–2394.

Donath, Judith, and danah boyd. "Public Displays of Connection." *BT Technology Journal* 22 (4) (2004): 71–82.

Donner, Jonathan. *After Access: Inclusion, Development, and a More Mobile Internet*. Cambridge, MA: MIT Press, 2015.

Douglas, Susan J. *Listening In: Radio and the American Imagination*. New York: Times Books, 1999.

Duffy, Brooke Erin. *(Not) Getting Paid to Do What You Love*. New Haven, CT: Yale University Press, 2017.

Farman, Jason. *Mobile Interface Theory: Embodied Space and Locative Media*. New York: Routledge, 2012.

Federici, Silvia. *Wages against Housework*. Bristol, UK: Power of Women Collective and Falling Wall Press, 1975.

Fischer, Claude. *America Calling: A Social History of the Telephone to 1940*. Berkeley: University of California Press, 1992.

Flanagan, Linda. "Reflecting on Adolescence: How Stories Can Inspire Teen Empathy." KQED News, June 15, 2016. https://kqed.org/mindshift/2016/06/15/mortified-how-sharing-childhood-journals-can-help-teens-feel-more-understood/.

Fothergill, Robert A. "One Day at a Time: The Diary as Lifewriting." *a/b: Auto/Biography Studies* 10 (1) (1995): 81–91.

Fothergill, Robert A. *Private Chronicles: A Study of English Diaries*. London: Oxford University Press, 1974.

Foucault, Michel. "Technologies of the Self." In *Technologies of the Self: A Seminar with Michel Foucault*, edited by Luther H. Martin, Huck Gutman, and Patrick H. Hutton, 16–49. Amherst: University of Massachusetts Press, 1988.

Fraser, Nancy. "Rethinking the Public Sphere." In *Habermas and the Public Sphere*, edited by Craig Calhoun, 109–142. Cambridge, MA: MIT Press, 1992.

Frosh, Paul. "The Gestural Image: The Selfie, Photography Theory, and Kinesthetic Sociability." *International Journal of Communication* 9 (2015): 1607–1628. http://ijoc.org/index.php/ijoc/article/view/3146/1388/.

Fuchs, Christian. "Digital Prosumption Labour on Social Media in the Context of the Capitalist Regime of Time." *Time & Society* 23 (1) (2014): 97–123.

Garfinkel, Simson, and David Cox. "Finding and Archiving the Internet Footprint." Paper presented at the First Digital Lives Research Conference: Personal Digital Archives for the 21st Century, London, UK, February 9–11, 2009.

Garvey, Ellen Gruber. *The Adman in the Parlor: Magazines and the Gendering of Consumer Culture, 1880s to 1910s*. New York: Oxford University Press, 1996.

Garvey, Ellen Gruber. *Writing with Scissors: American Scrapbooks from the Civil War to the Harlem Renaissance*. New York: Oxford University Press, 2012.

Gaver, William W. "Technology Affordances." Paper presented at the Association for Computing Machinery SIGCHI Conference on Human Factors in Computing Systems: Reaching through Technology, New Orleans, LA, April 27–May 2, 1991.

Gay, Geri. *Context-Aware Mobile Computing: Affordances of Space, Social Awareness, and Social Influence*. San Rafael, CA: Morgan and Claypool, 2009.

Geiger, Stuart R. "Bots, Bespoke, Code and the Materiality of Software Platforms." *Information Communication and Society* 17 (3) (2014): 342–356.

Gilbert, Eric, Saeideh Bakhshi, Shuo Chang, and Loren Terveen. "'I Need to Try This!': A Statistical Overview of Pinterest." Paper presented at the Association for Computing Machinery SIGCHI Conference on Human Factors in Computing Systems, Paris, France, April 27–May 2, 2013.

Gillespie, Tarleton. "The Relevance of Algorithms." In *Media Technologies: Essays on Communication, Materiality, and Society*, edited by Tarleton Gillespie, Pablo Boczkowski, and Kirsten A. Foot, 167–193. Cambridge, MA: MIT Press, 2014.

Gillespie, Tarleton, Pablo J. Boczkowski, and Kirsten A. Foot. "Introduction." In *Media Technologies: Essays on Communication, Materiality, and Society*, edited by Tarleton Gillespie, Pablo Boczkowski, and Kirsten A. Foot, 1–17. Cambridge, MA: MIT Press, 2014.

Gitelman, Lisa. *Always Already New: Media, History, and the Data of Culture*. Cambridge, MA: MIT Press, 2006.

Gitelman, Lisa. *Paper Knowledge: Toward a Media History of Documents*. Durham, NC: Duke University Press, 2014.

Glaser, Britney. "Mother Will Continue Sharing Photos of Her Stillborn Son, Despite Backlash." KPLC, October 2, 2015. http://www.kplctv.com/story/30164909/mother-will-continue-sharing-photos-of-her-stillborn-son-despite-backlash/.

Goffman, Erving. *Interaction Ritual: Essays on Face-to-Face Behavior*. Chicago: Aldine Publishers, 1967.

Goffman, Erving. *The Presentation of Self in Everyday Life*. Garden City, NY: Doubleday Anchor Books, 1959.

Gold, Katherine J., Vanessa K. Dalton, and Thomas L. Schwenk. "Hospital Care for Parents after Perinatal Death." *Obstetrics and Gynecology* 109 (5) (2007): 1156–1166.

Golden, Janet, and Lynn Weiner. "Reading Baby Books: Medicine, Marketing, Money and the Lives of American Infants." *Journal of Social History* 44 (3) (2011): 667–687.

Good, Katie Day. "From Scrapbook to Facebook: A History of Personal Media Assemblage and Archives." *New Media & Society* 15 (4) (2013): 557–573.

GoPro. "2015 Annual Financial Report." 2016. https://s21.q4cdn.com/291350743/files/doc_financials/2015/GoPro_-_2015_Annual_Report.pdf.

"GoPro Awards: 14-Year-Old Shreds Chopin Piano Solo in 4K." YouTube video, 1:10. Posted by "GoPro." April 11, 2016. https://youtube.com/watch?v=AyuAHa3JVfc.

Graves, Lucas. "The Affordances of Blogging: A Case Study in Culture and Technological Effects." *Journal of Communication Inquiry* 31 (4) (2007): 331–346.

Greenberg, Bradley S., Matthew S. Eastin, Paul Skalski, Len Cooper, Mark Levy, and Ken Lachlan. "Comparing Survey and Diary Measures of Internet and Traditional Media Use." *Communication Reports* 18 (1) (2005): 1–8.

Grice, H. P. "Logic in Conversation." In *Syntax and Semantics, Vol. 3: Speech Acts*, edited by Peter Cole and Jerry L. Morgan, 41–58. New York: Academic Press, 1972.

Griffith, Maggie, and Zizi Papacharissi. "Looking for You: An Analysis of Video Blogs." *First Monday* 15 (1–4) (2009). http://firstmonday.org/ojs/index.php/fm/article/view/2769/2430/.

Habermas, Jürgen. *The Structural Transformation of the Public Sphere: An Inquiry into a Category of Bourgeois Society*. Translated by Thomas Burger. Cambridge, MA: MIT Press, 1962.

Halpern, Megan, and Lee Humphreys. "Iphoneography as an Emergent Art World." *New Media & Society* 18 (1) (2016): 62–81.

Hardt, Michael. "Affective Labor." *boundary 2* 26 (2) (1999): 89–100.

Harvey, David. *The Condition of Postmodernity*. Malden, MA: Wiley-Blackwell, 1990.

Hearn, Alison "Structuring Feeling: Web 2.0, Online Ranking and Rating, and the Digital 'Reputation' Economy." *ephemera: theory & politics in organizations* 10 (3–4) (2010): 421–438.

Heehs, Peter. *Writing the Self: Diaries, Memoirs, and the History of the Self.* New York: Bloomsbury Publishing USA, 2013.

Helsen, Werner F, and Janet L. Starkes. "A Multidimensional Approach to Skilled Perception and Performance in Sport." *Applied Cognitive Psychology* 13 (1) (1999): 1–27.

Henkin, David M. *The Postal Age: The Emergence of Modern Communications in Nineteenth-Century America.* Chicago: University of Chicago Press, 2007.

Hesmondhalgh, David, and Sarah Baker. *Creative Labour: Media Work in Three Cultural Industries.* London: Routledge, 2013.

Highfield, Tim. "News via Voldemort: Parody Accounts in Topical Discussions on Twitter." *New Media & Society* 18 (9) (2015): 2028–2045.

Highfield, Tim, Stephen Harrington, and Axel Bruns. "Twitter as a Technology for Audiencing and Fandom." *Information Communication and Society* 16 (3) (2013): 315–339.

Highmore, Ben. *Ordinary Lives: Studies in the Everyday.* London: Routledge, 2010.

Hilliker, Laurel. "Letting Go While Holding On: Postmortem Photography as an Aid in the Grieving Process." *Illness, Crisis & Loss* 14 (3) (2006): 245–269.

Hillis, Ken. *Online a Lot of the Time: Ritual, Fetish, Sign.* Durham, NC: Duke University Press, 2009.

Hirsch, Julie. *Family Photographs: Content, Meaning, and Effect.* New York: Oxford University Press, 1981.

Hjorth, Larissa, and Sun Sun Lim. "Mobile Intimacy in an Age of Affective Mobile Media." *Feminist Media Studies* 12 (4) (2012): 477–484.

Hogan, Bernie. "The Presentation of Self in the Age of Social Media: Distinguishing Performances and Exhibitions Online." *Bulletin of Science, Technology & Society* 30 (6) (2010): 377–386.

Horkheimer, Max, and Theodor W. Adorno. "The Culture Industry: Enlightenment as Mass Deception." In *Dialectic of Enlightenment,* edited by Max Horkheimer and Theodor W. Adorno, 120–167. New York: Continuum, 1998.

Hoschild, Arlie, and Anne Machung. *The Second Shift: Working Parents and the Revolution at Home.* New York: Viking, 1989.

Howard, Philip N., and Muzammil M. Hussain. *Democracy's Fourth Wave?: Digital Media and the Arab Spring.* Oxford: Oxford University Press, 2013.

Humphreys, Ashlee. *Social Media: Enduring Principles*. New York: Oxford University Press, 2015.

Humphreys, Lee. "Technological Determinism." In *Encyclopedia of Science and Technology Communication*, edited by Susanna Hornig Priest, 869–872. Thousand Oaks, CA: Sage Publications, 2010.

Humphreys, Lee. "Who's Watching Whom? A Study of Interactive Technology and Surveillance." *Journal of Communication* 61 (4) (2011): 575–595. doi:10.1111/j.1460-2466.2011.01570.x.

Humphreys, Lee, Phillipa Gill, Balachander Krishnamurthy, and Elizabeth Newbury. "Historicizing New Media: A Content Analysis of Twitter." *Journal of Communication* 63 (3) (2013): 413–431.

Humphreys, Lee, Thilo von Pape, and Veronika Karnowski. "Evolving Mobile Media: Uses and Conceptualizations of the Mobile Internet." *Journal of Computer-Mediated Communication* 18 (4) (2013): 491–507.

Hunter, Jane H. "Inscribing the Self in the Heart of the Family: Diaries and Girlhood in Late-Victorian America." *American Quarterly* 44 (1) (1992): 51–81.

Huppke, Rex W. "A Tower of Babble, Built out of Old Tweets." *Chicago Tribune*, January 8, 2013.

Innis, Harold A. "The Bias of Communication." *Canadian Journal of Economics and Political Science* 15 (4) (1949): 457–476.

Ito, Mizuko. "Introduction: Personal, Portable, Pedestrian." In *Personal, Portable, Pedestrian: Mobile Phones in Japanese Life*, edited by Mizuko Ito, Daisuke Okabe, and Misa Matsuda, 1–16. Cambridge, MA: MIT Press, 2005.

Jackson, Heather J. *Marginalia: Readers Writing in Books*. New Haven, CT: Yale University Press, 2002.

Jenkins, Reese. *Images and Enterprise: Technology and the American Photographic Industry, 1839 to 1925*. Baltimore: Johns Hopkins University Press, 1975.

Jenkins, Reese V. "Technology and the Market: George Eastman and the Origins of Mass Amateur Photography." *Technology and Culture* 16 (1) (1975): 1–19.

Jensen, Klaus Bruhn. Media. In *International Encyclopedia of Communication*, ed. Wolfgang Donsbach. Blackwell Publishing, 2008., 10.1111/b.9781405131995.2008.x.

Jervis, Rick. "Disney Joins Growing Number of Venues Banning Selfie Sticks." *USA Today*, June 26, 2015. http://www.usatoday.com/story/news/2015/06/26/disney-selfie-stick-ban/29333857/.

Johnson, Gabe, and Nick Wingfield. "GoPro Goes Amateur." *New York Times*, January 31, 2014. https://www.nytimes.com/2014/01/31/technology/gopro-works-on-its-brand.html.

Katcher, Allan. "Self-Fulfilling Prophecies and Active Listening." In *Bridges Not Walls: A Book about Interpersonal Communication*, edited by John Stewart, 81–88. Reading, MA: Addison-Wesley, 1973.

Keegan, Amanda. "4 Ways to Save Money on Your Cell Phone Bill." ABC News, May 26, 2015. http://abcnews.go.com/Business/ways-save-money-cell-phone-bill/story?id=31297718.

Kelley, Steve. "It's a tweet from Amanda…" *Times Picayune*, April 9, 2009. http://editorialcartoonists.com/cartoon/display.cfm/69138/.

King, Graham. *Say "Cheese"! Looking at Snapshots in a New Way*. New York: Dodd, Mead and Company, 1984.

Koerber, Brian. "An Unofficial Guide for #ThrowbackThursday Etiquette." *Mashable*, April 17, 2014. http://mashable.com/2014/04/17/throwback-thursday-etiquette/.

Kubey, Robert, Reed Larson, and Mihaly Czikszentmihalyi. "Experience Sampling Method Applications to Communication Research Questions." *Journal of Communication* 4 (3) (1996): 99–120.

Kuhn, Annette. "A Journey through Memory." In *Memory and Methodology*, edited by Susannah Radstone, 179–196. Oxford: Berg, 2000.

Kuhn, Annette. "Memory Texts and Memory Work: Performances of Memory in and with Visual Media." *Memory Studies* 3 (4) (2010): 298–313.

Lambert, Ronald D. "The Family Historian and Temporal Orientations towards the Ancestral Past." *Time & Society* 5 (2) (1996): 115–143.

Lasch, Christopher. *The Culture of Narcissism: American Life in an Age of Diminishing Expectations*. New York: Warner Books, 1979.

Lasen, Amparo. "Affective Technologies—Emotions and Mobile Phones." *Receiver* 11 (2004): 1–8.

Lepore, Jill. *Book of Ages: The Life and Opinions of Jane Franklin*. New York: Vintage Books, 2014.

Lievrouw, Leah. "New Media, Mediation, and Communication Study." *Information Communication and Society* 12 (3) (2009): 303–325.

Litt, Eden. "Knock, Knock. Who's There? The Imagined Audience." *Journal of Broadcasting & Electronic Media* 56 (3) (2012): 330–345.

Lohmeier, Christine, and Christian Pentzold. "Making Mediated Memory Work: Cuban-Americans, Miami Media and the Doings of Diaspora Memories." *Media, Culture & Society* 36 (6) (2014): 776–789.

Lupton, Deborah. *The Quantified Self*. London: Polity, 2016.

Marvin, Carolyn. *When Old Technologies Were New: Thinking about Electric Communication in the Late Nineteenth Century*. New York: Oxford University Press, 1988.

Marwick, Alice E. *Status Update: Celebrity and Attention in the Social Media Age*. New Haven, CT: Yale University Press, 2013.

Marwick, Alice E., and danah boyd. "I Tweet Honestly, I Tweet Passionately: Twitter Users, Context Collapse, and the Imagined Audience." *New Media & Society* 13 (1) (2011): 114–133.

Matthews, Nicole. "Confessions to a New Public: Video Nation Shorts." *Media, Culture & Society* 29 (3) (2007): 435–448.

Mayer, Michael, dir. *Mortified Nation*. [Documentary] Los Angeles: Wiser Post, 2013.

Mayer-Schönberger, Viktor, and Kenneth Cukier. *Big Data: A Revolution That Will Transform How We Live, Work, and Think*. Boston: Houghton Mifflin Harcourt, 2013.

McCarthy, Molly. "A Pocketful of Days: Pocket Diaries and Daily Record Keeping among Nineteenth-Century New England Women." *New England Quarterly* 73 (2) (2000): 274–296.

McCarthy, Molly A. *The Accidental Diarist: A History of the Daily Planner in America*. Chicago: University of Chicago Press, 2013.

McCrum, Sean. "Snapshot Photography." *Circa* 60 (November–December) (1991): 32–36.

McGee, Kevin, and Jörgen Skågeby. "Gifting Technologies." *First Monday* 9 (12) (2004). http://firstmonday.org/ojs/index.php/fm/article/view/1192/1112/.

McGuinness, Denise, Barbara Coughlan, and Sheila Power. "Empty Arms: Supporting Bereaved Mothers during the Immediate Postnatal Period." *British Journal of Midwifery* 22 (4) (2014): 246–252.

Mead, George Herbert. *Mind, Self a Society from the Standpoint of a Social Behaviorist*. Chicago: University of Chicago Press, 1934.

Meese, James, Martin Gibbs, Marcus Carter, Michael Arnold, Bjorn Nansen, and Tamara Kohn. "Selfies at Funerals: Mourning and Presencing on Social Media Platforms." *International Journal of Communication* 9 (2015): 1818–1831.

Messaris, Paul. *Visual "Literacy": Image, Mind, and Reality*. Boulder, CO: Westview Press, 1994.

Meyer, Eric. "Inadvertent Algorithmic Cruelty." Personal website. December 24, 2014. http://meyerweb.com/eric/thoughts/2014/12/24/inadvertent-algorithmic-cruelty/.

Meyrowitz, Joshua. *No Sense of Place: The Impact of Electronic Media on Social Behavior*. New York: Oxford University Press, 1985.

Meyrowitz, Joshua. "Watching Us Being Watched: State, Corporate, and Citizen Surveillance." Paper presented at the symposium *The End of Television? Its Impact on the World (So Far)*. Annenberg School for Communication, University of Pennsylvania, Philadelphia, February 17–18, 2007.

Miller, Susan. *Assuming the Position: Cultural Pedagogy and the Politics of Commonplace Writing*. Pittsburgh, PA: University of Pittsburgh Press, 1998.

Milne, Esther. *Letters, Postcards, Email: Technologies of Presence*. New York: Routledge, 2010.

Moller, Kenza. "Why You Shouldn't Post Photos of 9/11 on Social Media." Romper, September 11, 2016. https://www.romper.com/p/why-you-shouldnt-post-photos-from-911-on-social-media-18144/.

Moreau, Elise. "What Throwback Thursday Actually Is and Why It's So Popular." *Life Wire*, August 14, 2017. https://www.lifewire.com/throwback-thursday-meaning-and-why-its-so-popular-3485860/.

Murphy, Kate. "Blogging with Video, Hoping to Go Viral." *New York Times*, February 14, 2013. http://www.nytimes.com/2013/02/14/technology/personaltech/how-to-make-your-video-go-viral.html.

Naaman, Mor, Jeffrey Boase, and Chih-Hui Lai. "Is It Really About Me? Message Content in Social Awareness Streams." *Proceedings of the ACM Conference on Computer Supported Collaborative Work (CSCW2010)*, 189–192. New York: Association for Computing Machinery, 2010. doi:10.1145/1718918.1718953.

Nahon, Karine, and Jeff Hemsley. *Going Viral*. Cambridge: Polity, 2013.

Neff, Gina, and Dawn Nafus. *Self-Tracking*. Cambridge, MA: MIT Press, 2016.

Nemer, David, and Guo Freeman. "Empowering the Marginalized: Rethinking Selfies in the Slums of Brazil." *International Journal of Communication* 9 (2015): 1832–1847.

Novielli, Carole. "Remembering Grayson: Anencephalic Baby Facebook Banned Whose Life Impacted So Many." Life News, March 3, 2014. http://www.lifenews.com/2014/03/03/remembering-grayson-anencephalic-baby-facebook-banned-whose-life-impacted-so-many/.

O'Keeffe, Gwenn Schurgin, and Kathleen Clarke-Pearson. "The Impact of Social Media on Children, Adolescents, and Families." *Pediatrics* 127 (4) (2011): 800–804.

Oakley, Ann. *Woman's Work: The Housewife, Past and Present.* New York: Vintage, 1974.

Odih, Pamela. "Gendered Time in the Age of Deconstruction." *Time & Society* 8 (1) (1999): 9–38.

Ong, Walter. *Orality and Literacy*. London: Routledge, 2012.

Ott, Katherine, Susan Tucker, and Patricia P. Buckler. "An Introduction to the History of Scrapbooks." In *The Scrapbook in American Life,* edited by Susan Tucker, Katherine Ott, and Patricia P. Buckler, 1–25. Philadelphia, PA: Temple University Press, 2006.

Papacharissi, Zizi. *Affective Publics: Sentiment, Technology, and Politics*. Oxford: Oxford University Press, 2015.

Papacharissi, Zizi. "The Virtual Sphere 2.0: The Internet, the Public Sphere, and Beyond." In *Handbook of Internet Politics,* edited by Andrew Chadwick and Phil Howard, 230–245. Abingdon, UK: Routledge, 2008.

Pariser, Eli. *The Filter Bubble: How the New Personalized Web Is Changing What We Read and How We Think*. London: Penguin Books, 2012.

Payne, Valerie. "Grieving Mom Refuses to Remove Photos of Stillborn Son from Facebook." HLN, October 5, 2015. http://www.hlntv.com/articles/2015/10/05/mom-posts-photos-of-stillborn-son-on-facebook-sparks-controversy/.

Peirce, Charles Sanders. *Peirce on Signs: Writings on Semiotic by Charles Sanders Peirce.* Edited by James Hoopes. Chapel Hill: University of North Carolina Press, 1991.

Peters, Mitchell. "Music Stars Remember 15th Anniversary of 9/11 on Social Media." *Billboard*, September 11, 2016. http://www.billboard.com/articles/news/7503633/music-stars-remember-15th-anniversary-of-911-september-11-never-forget-world-trade-center/.

Petrow, Steven. "Navigating the Rules of 'Throwback Thursday.'" *USA Today*, August 20, 2015. https://www.usatoday.com/story/tech/personal/2015/08/20/navigating-rules-throwback-thursday/32041277/.

Pinch, Trevor J., and Wiebe E. Bijker. "The Social Construction of Facts and Artefacts: Or How the Sociology of Science and the Sociology of Technology Might Benefit Each Other." *Social Studies of Science* 14 (3) (1984): 399–441.

Pooley, Jefferson. "The Consuming Self: From Flappers to Facebook." In *Blowing up the Brand: Critical Perspectives on Promotional Culture,* edited by Melissa Aronczyk and Devon Powers, 71–89. New York: Peter Lang, 2010.

Poster, Mark. *The Mode of Information: Poststructuralism and Social Construct.* Chicago: University of Chicago Press, 1990.

Rakow, Lana F., and Laura A. Wackwitz, eds. *Feminist Communication Theory: Selections in Context.* Thousand Oaks, CA: Sage, 2004.

Raymond, Matt. "How Tweet It Is!: Library Acquires Entire Twitter Archive." Library of Congress, April 14, 2010. http://blogs.loc.gov/loc/2010/04/how-tweet-it-is-library-acquires-entire-twitter-archive/.

Reinecke, Leonard, and Sabine Trepte. "Authenticity and Well-Being on Social Network Sites: A Two-Wave Longitudinal Study on the Effects of Online Authenticity and the Positivity Bias in SNS Communication." *Computers in Human Behavior* 30 (2014): 95–102.

Rettberg, Jill Walker. *Seeing Ourselves through Technology: How We Use Selfies, Blogs and Wearable Devices to See and Shape Ourselves.* Basingstoke, UK: Palgrave Macmillan, 2014.

Richardson, Ingrid, and Rowan Wilken. "Parerga of the Third Screen: Mobile Media, Place, Presence." In *Mobile Technology and Place,* edited by Rowan Wilken and Gerard Goggin, 181–197. New York: Routledge, 2012.

Sampson, Edward E. "The Deconstruction of the Self." In *Texts of Identity,* edited by John Shotter and Kenneth J. Gergen, 1–19. London: Sage, 1989.

Schrader, Alex. "Teens, Social Media, and Narcissism." Edudemic, January 30, 2017. http://www.edudemic.com/teens-social-media-narcissism/.

Schrock, Andrew Richard. "Communicative Affordances of Mobile Media: Portability, Availability, Locatability, and Multimediality." *International of Communication* 9 (2015): 1235–1238. http://ijoc.org/index.php/ijoc/article/view/3288/1363/.

Scolere, Leah, and Lee Humphreys. "Pinning Design: The Curatorial Labor of Creative Professionals." *Social Media + Society* 2 (1) (2016). doi:10.1177/2056305116633481.

Senft, Theresa M. *Camgirls: Celebrity and Community in the Age of Social Networks.* New York: Peter Lang, 2008.

Senft, Theresa M., and Nancy K. Baym. "What Does the Selfie Say? Investigating a Global Phenomenon." *International Journal of Communication* 9 (2015): 1588–1606. http://ijoc.org/index.php/ijoc/article/view/4067/1387/.

Sennett, Richard. *The Fall of Public Man.* London: Norton, 1976.

Serota, Herman M. "Home Movies of Early Childhood: Correlative Developmental Data in the Psychoanalysis of Adults." *Science* 143 (3611) (1964): 1195.

Shifman, Limor. *Memes in Digital Culture.* Cambridge, MA: MIT Press, 2014.

Silverstone, Roger. "The Sociology of Mediation and Communication." In *The Sage Handbook of Sociology,* edited by Craig Calhoun, Chris Rojek, and Bryan Turner, 188–207. London: Sage, 2005.

Silverstone, Roger, and Leslie Haddon. "Design and the Domestication of Information and Communication Technologies: Technical Change and Everyday Life." In *Communication by Design. The Politics of Information and Communication Technologies*, edited by Robin Mansel and Roger Silverstone, 44–74. Oxford: Oxford University Press, 1996.

Sontag, Susan. *On Photography*. New York: Anchor Books Doubleday, 1989.

Spigel, Lynn. *Make Room for TV: Television and the Family Ideal in Postwar America*. Chicago: University of Chicago Press, 1992.

Stark, David. *The Sense of Dissonance: Accounts of Worth in Economic Life*. Princeton, NJ: Princeton University Press, 2009.

Stein, Joel. "Millennials: The Me Me Me Generation." *Time*, May 9, 2015. http://time.com/247/millennials-the-me-me-me-generation/.

Steinitz, Rebecca. "Writing Diaries, Reading Diaries: The Mechanics of Memory." *Communication Review* 2 (1) (1997): 43–58.

Sterne, Jonathan. "Analog." In *Digital Keywords: A Vocabulary of Information Society and Culture*, edited by Benjamin Peters, 31–44. Princeton, NJ: Princeton University Press, 2016.

Taffel, Sy. "Perspectives on the Postdigital: Beyond Rhetorics of Progress and Novelty." *Convergence* 22 (3) (2016): 324–338.

Terranova, Tiziana. "Free Labor: Producing Culture for the Digital Economy." *Social Text* 18 (2) (2000): 33–58.

Thelwall, Mike, Kevan Buckley, and Georgios Paltoglou. "Sentiment in Twitter Events." *Journal of the American Society for Information Science and Technology* 62 (2) (2011): 406–418.

Thumim, Nancy. *Self-Representation and Digital Culture*. Basingstoke, UK: Palgrave Macmillan, 2012.

Tsukayama, Hayley. "Fueled by Stoke, GoPro Helps Turn Rookie Selfie Snappers into Film Directors." *Washington Post*, October 10, 2014. http://wapo.st/1wb0qSu/.

Tucker, Susan, Katherine Ott, and Patricia P. Buckler, eds. *The Scrapbook in American Life*. Philadelphia, PA: Temple University Press, 2006.

Tufekci, Zeynep, and Christopher Wilson. "Social Media and the Decision to Participate in Political Protest: Observations from Tahrir Square." *Journal of Communication* 62 (2) (2012): 363–379.

Turkle, Sherry. *Alone Together: Why We Expect More from Technology and Less from Each Other*. New York: Basic Books, 2011.

Turow, Joseph. *The Aisles Have Eyes: How Retailers Track Your Shopping, Strip Your Privacy, and Define Your Power*. New Haven, CT: Yale University Press, 2017.

Turow, Joseph. *The Daily You: How the New Advertising Industry Is Defining Your Identity and Your Worth*. New Haven, CT: Yale University Press, 2011.

Twitter, Inc. "2016 Annual Report." February 27, 2017. http://bit.ly/2yCzZ0g/.

van Dijck, José. "Datafication, Dataism and Dataveillance: Big Data between Scientific Paradigm and Ideology." *Surveillance & Society* 12 (2) (2014): 197–208.

van Dijck, José. "Flickr and the Culture of Connectivity: Sharing Views, Experiences, Memories." *Memory Studies* 4 (4) (2010): 401–415.

van Dijck, José. *Mediated Memories in the Digital Age*. Stanford, CA: Stanford University Press, 2007.

van Dijck, José. "'You Have One Identity': Performing the Self on Facebook and LinkedIn." *Media, Culture & Society* 35 (2) (2013): 199–215.

van Zoonen, Liesbet, and Georgina Turner. "Exercising Identity: Agency and Narrative in Identity Management." *Kybernetes* 43 (6) (2014): 935–946.

Varnelis, Kazys, ed. *Networked Publics*. Cambridge, MA: MIT Press, 2008.

Veblen, Thorstein. *The Theory of the Leisure Class: An Economic Study of Institutions*. New York: Modern Library, 1934.

Vitak, Jessica, Cliff Lampe, Rebecca Gray, and Nicole B. Ellison. "'Why Won't You Be My Facebook Friend?': Strategies for Managing Context Collapse in the Workplace." Paper presented at the 2012 iConference, Toronto, Ontario, Canada, February 7–10, 2012.

Wajcman, Judy. *Pressed for Time: The Acceleration of Life in Digital Capitalism*. Chicago: University of Chicago Press, 2015.

Walkley, Ellen. "OHQ Research Files: Scraps of History: Researching Scrapbooks at the Oregon Historical Society." *Oregon Historical Quarterly* 102 (4) (2001): 512–517.

West, Nancy Martha. *Kodak and the Lens of Nostalgia*. Charlottesville: University of Virginia, 2000.

"What Is Snapchat?" YouTube video, 3:56. Posted by "Snapchat." June 16, 2015. https://youtu.be/ykGXIQAHLnA.

Williams, Raymond. *Culture and Society*. New York: Harper and Row, 1966.

Williams, Raymond. *Television: Technology and Cultural Form*. New York: Schocken Books, 1975.

Wolcott, Harry F. *Transforming Qualitative Data: Description, Analysis, and Interpretation*. Thousand Oaks, CA: Sage, 1994.

Yakel, Elizabeth. "Seeking Information, Seeking Connections, Seeking Meaning: Genealogists and Family Historians." *Information Research: An International Electronic Journal* 10 (1) (2004). http://www.informationr.net/ir/10-1/paper205.html.

Zelizer, Barbie. *Remembering to Forget: Holocaust Memory through the Camera's Eye*. Chicago: University of Chicago Press, 1998.

Zittrain, Jonathan. *The Future of the Internet: And How to Stop It*. New Haven, CT: Yale University Press, 2008.

Zuromskis, Catherine. *Snapshot Photography: The Lives of Images*. Cambridge, MA: MIT Press, 2013.

Index

Page references in italics indicate illustrations.

Printed in the United States
by Baker & Taylor Publisher Services